Dreams of Calculus

Johan Hoffman
Claes Johnson
Anders Logg

Dreams of Calculus

Perspectives on Mathematics Education

 Springer

Johan Hoffman
Courant Institute
of Mathematical Sciences
New York University
251 Mercer Street
New York, NY 10012-1185, USA
e-mail: hoffman@cims.nyu.edu

Anders Logg
Department of Computational
Mathematics
Chalmers University
of Technology
412 96 Göteborg, Sweden
e-mail: logg@math.chalmers.se

Claes Johnson
Department of Computational
Mathematics
Chalmers University
of Technology
412 96 Göteborg, Sweden
e-mail: claes@math.chalmers.se

Library of Congress Control Number: 2004108047

Mathematics Subject Classification (2000): 00A05, 00A06, 00A30, 00A69, 00A71, 00A72, 00A99, 01-08, 65L60, 65M60

ISBN 3-540-21976-5 Springer Berlin Heidelberg New York

Springer is a part of Springer Science+Business Media
springeronline.com
© Springer Berlin Heidelberg 2004
Printed in Germany

The use of general descriptive names, registered names, trademarks, etc. in this publication does not imply, even in the absence of a specific statement, that such names are exempt from the relevant protective laws and regulations and therefore free for general use.

Typeset by the authors
using a Springer TEX macro package

Cover design: *design & production*
Printed on acid-free paper   46/3142LK - 5 4 3 2 1 0

This book is dedicated to the Swedish Minister of Education Thomas Östros, who created the Mathematics Delegation on January 23 2003.

Preface

The purpose of this book is to stimulate a much needed debate on mathematics education reform. We do not claim that we present anything near the whole truth, and that there are not many other points of view. We thus invite to debate and urge other people to express their views. As scientists we believe that it is our duty to present our own standpoints and conclusions as clearly as possible, to open for scrutiny and discussion, rather than ambiguous politically correct views, which are difficult to question.

This book may be viewed as an introduction to our mathematics education reform project Body&Soul based on a synthesis of Body (computational mathematics) and Soul (analytical mathematics). We do make a case for computational mathematics, which is the new world of mathematics opened by the computer, now waiting to enter mathematics education. But we do also propose a synthesis of this new world, with the kernel being computational algorithms, and the traditional world of analytical mathematics, with analytical formulas as the kernel. In fact, algorithms are expressed using analytical formulas before being translated into computer code, and good analytical formulas are necessary for understanding and insight.

Reviewers of the book have remarked that what we tell about the situation in Sweden is hard to believe, and should better be omitted. Yet, what we tell is indeed true, and following the idea that true observations of individual facts often may point to a universal truth, while speculations without observations often lead nowhere, we have decided to tell (a bit of) the story. And we believe that after all Sweden is not that special, maybe even typical.

We hope the book can be read (and understood) by many. We hope the reader will find that the book is partly serious, partly not so serious, and that our prime objective is to be constructive and contribute something new, not to be critical of traditional views. Finally, we hope the reader will accept that the book is a sketch (not a treatise), which was written quickly, and can also be read quickly.

Göteborg in April 2004

Johan Hoffman, Claes Johnson and Anders Logg

Contents

summary of their thesis ... 138

III   Appendix   **141**

Part I

Perspectives

1

Introduction

21 Oct 2005

...the resolution of revolutions is the selection by conflict within
the scientific community of the fittest way to practice further science... (Thomas Kuhn, *The Structure of Scientific Revolutions*)

1.1 The Mathematics Delegation and its Main Task

The *Mathematics Delegation* was created by the Minister of Education
Thomas Östros at a Swedish government cabinet meeting on 23rd January
2003 with the following *main task*:

- *Analyse the current situation in terms of the teaching of mathematics
  in Sweden and assess the need for changing current syllabuses and
  other steering documents.*

The report of the delegation is to be delivered on May 28 2004. The delegation involves more than 100 people in several different committees, but
contains only one professor of mathematics, and in particular no expertise
of computational mathematics.

1.2   Crisis and Change of Paradigm, or Not?

In public presentations of the work of the Mathematics Delegation by its chairman, the following two statements have been expressed as a basis for the analysis:

- A1: *There is no crisis* in the mathematics education today.

- A2: *There is no change of paradigm* in mathematics education now going on because of the computer.

Our motivation to write this book comes from our conviction that

- B1: *There is a crisis* in the mathematics education today.

- B2: *There is a change of paradigm* in mathematics education now going on because of the computer.

We base our conviction on our work as experts in the field of computational mathematics. In particular, we believe that the present crisis is caused by an ongoing change of paradigm.

To decide whether A1–2 or B1–2 is true (they cannot be true both) is important: If the analysis of the delegation is based on A1–2, and indeed B1–2 is true and not A1–2, it would seem to be a high risk that the main task of the delegation cannot be met. And this task is very important, since mathematics education touches so many in the educational system, and has such an important role in our society.

We encourage the reader to seek to take a standpoint. In this book, we give our arguments in favor of B1–2 in a form which hopefully is readable to many.

1.3   The Body&Soul Project

Of course, the problems are not solved by just identifying B1–2 as more true than A1–2; if B1–2 indeed describe the realities, it remains to come up with a reformed mathematics program reflecting the change of paradigm and which may help to resolve the crisis. We briefly present our work in this direction within our Body&Soul mathematics education reform project (see www.phi.chalmers.se/bodysoul/).

The Body&Soul project has grown out of a 30 year activity of the senior author in international research with several influential articles and books. This book may be viewed as a kind of summary of this work, accomplished with the help of several younger coworkers now carrying the project further.

1.4   Same Questions in All Countries

Mathematics education is of course not a Swedish affair; the issues are similar in all countries, and in each country the different points of view with A1–2 and B1–2 as clearly expressed alternatives, have their supporters.

The book is thus directed to a large audience also outside Sweden, although the direct stimuli to write the book as indicated came from the Mathematics Delegation in Sweden.

If not already in existence, there may soon be a Mathematics Delegation in many countries.

1.5   Why We Wrote this Book

In our contacts with the Mathematics Delegation and also elsewhere, we have met attitudes indicating that a common ground for discussion on issues of mathematics education is largely lacking. It appears that today the worlds of pure mathematics, computational mathematics and mathematics didactics are largely separated with little interaction. The result is that there are today no clear answers to the basic questions which are generally accepted and acknowledged.

The motivation to write this book comes from this confusion. We hope to be able to exhibit some key aspects of mathematics education today and present some constructive elements to help create a more fruitful climate for debate and reform. Of course, we feel that this is not an easy task.

We hope this book will be readable for a general audience without expert knowledge of mathematics.

To help, we use a minimum of mathematical language in the first part of the book, and try to express basic ideas and aspects in "just" words.

The second part is a bit more technical mathematical and contains some formulas. It may be used as a test of our claim that some mathematics is both understandable and useful.

2
What? How? For Whom? Why?

I admit that each and every thing remains in its state until there is reason for change. (Leibniz)

The mathematician's pattern's, like those of the painter's or the poet's, must be beautiful, the ideas, like the colours or the words, must fit together in a harmonious way. There is no permanent place in the world for ugly mathematics. (Hardy)

2.1 Mathematics and the Computer

Mathematics is an important part of our culture and has a central role in education from elementary pre-school, through primary, secondary schools and high schools to many university programs.

The *basic questions* of mathematics education are: *What to teach? How to teach? Whom to teach?* and *Why to teach?* The answers to these questions are changing over time, as mathematics as a science and our entire society is changing.

Education in general, and in particular education in Mathematics and Science, is supposed to have a scientific basis. A teacher claiming in class today that $2 + 2 = 5$, or that the Earth is flat, or denying the existence of electrons, bacteria or the genetic code, would certainly face severe difficulties.

The scientific basis of the standard mathematics education presented today was formed during the 19th century, well before the computer was invented starting in the mid 20th century. Therefore the current answers to the basic questions formulated above reflect a view of *mathematics without the computer*.

Today the computer is changing our lives and our society. The purpose of this book is twofold: First, we want give evidence that the computer is changing also mathematics as a science in a profound way, and thus new answers to the basic questions of mathematics education have to be given. Second, we want to propose new answers to these questions. We do not claim that our answers are the only possible, but we do claim that the old answers are no longer functional.

The purpose of this book is to stimulate a much needed debate on mathematics education. We seek to reach a wide public of teachers and students of mathematics on all levels, and we therefore seek to present some basic ideas as simply and clearly as possible, with a minimum of mathematical notation.

2.2   Pure and Computational Mathematics

The terms *pure mathematics* and *applied mathematics* are used to identify different areas of mathematics as a science, with different focus. In applied mathematics the main topics of investigation would come from areas such as mechanics and physics, while in pure mathematics one could pursue mathematical questions without any coupling to applications.

The distinction between pure and applied mathematics is quite recent and gradually developed during the 20th century. All the great mathematicians like Leibniz, Euler, Gauss, Cauchy, Poincare and Hilbert were generalists combining work in both pure and applied mathematics as well as mechanics and physics and other areas.

Even today, there is no clear distinction between pure and applied mathematics; a mathematical technique once developed within pure mathematics may later find applications and thus become a part of applied mathematics. Conversely, many questions within pure mathematics may be viewed as ultimately originating from applications.

Another distinction is now developing: *mathematics without computer* and *mathematics with computer*. Applied mathematics today can largely be described as mathematics with computer, or *computational mathematics*. Most of the activity of pure mathematics today can correspondingly be described as mathematics without computer, although computers have been used to solve some problems posed within pure mathematics. One example is the famous *4-color problem* asking for a mathematical proof that 4 colors are enough to color any map so that neighboring countries do not get the same color.

For the discussion below, we make the distinction between pure mathematics (essentially mathematics without computer), and computational mathematics (mathematics with computer).

The Information Society and Computational Mathematics

In fact, computational mathematics forms the basis of our new *information society* with digital word, sound and image in new forms of *virtual reality* within science, medicine, economy, education and entertainment. The scanner at the hospital giving detailed images of the interior of your body, the weather forecast, the flight simulator, the animated movie, the robot in the car factory, all use different forms of computational mathematics. We will return to some of these topics below.

Fermat's Last Theorem

In pure mathematics a question may receive attention just because it represents an intellectual challenge, not because it has a scientific relevance as a question of some importance to mankind. The famous mathematician G. Hardy (1877–1947) expressed this attitude very clearly in his book "A Mathematician's Apology", although the title indicates some doubts about public acceptance. The prime example of this form is the proof of the famous Fermat's Last Theorem in number theory, which was the most famous open problem in pure mathematics for 300 hundred years until Andrew Wiles completed his 130 page proof in 1994 after 8 years of heroic lonely constant struggle. For this achievement Wiles effectively received a Fields Medal in 1998, the Nobel Prize of Mathematics, at the International Congress of Mathematicians in Berlin 1998, although technically the Prize was awarded in the form of a Special Tribute connected to the Fields Medal awards (because Wiles had passed the limit of 40 years of age to receive the medal).

Fermat's Last Theorem states that there are no integers x, y and z which satisfy the equation $x^n + y^n = z^n$, where n is an integer larger than 2. It was stated in the notes of Fermat (1601–1665) in the margin of a copy of *Arithmetica* by Diofantes of Alexandria (around 250 AD). Fermat himself indicated a proof for $n = 4$ and Euler developed a similar proof for $n = 3$. The French Academy of Science offered in 1853 its big prize for a full proof and drew contributions from famous mathematicians like Cauchy, but none of the submitted proofs at the dead-line 1857 was correct (and none after that until Wiles proof in 1994). The great mathematician Gauss (1777–1855), called the king of mathematics, decided not to participate, because he viewed the problem to be of little interest. Gauss believed in a synthesis of pure and applied mathematics, with mathematics being the queen of science.

The form of Fermat's Last Theorem makes it particularly difficult to prove, since it concerns *non-existence* of integers x, y and z and an integer $n > 2$ such that $x^n + y^n = z^n$. The proof has to go by contradiction, by proving that the assumption of existence of a solution leads to a contradiction. It took Wiles 130 pages to construct a contradiction, in a proof which can be followed in detail by only a few true experts.

Fermat's Last Theorem may seem appealing to a pure mathematician because it is (very) easy to state, but (very) hard to prove. Thus, it can be posed to a large audience, but the secret of the solution is kept to a small group of specialists, as in the Pythagorean society built on number theory.

Gauss or Hardy?

But what is the scientific meaning of a proof of Fermat's Last Theorem? Some mathematicians may advocate that (apart from aesthetics) it is not the result itself that is of interest, but rather the methods developed to give a proof. Gauss would probably not be too convinced by this type of argument, unless some striking application was presented, while Hardy would be.

The incredible interest and attention that Wiles proof did draw within trend-setting circles of mathematics, shows that the point of view of Hardy, as opposed to that of Gauss, today is dominating large parts of the scene of mathematics, but the criticism of Gauss may still be relevant. We will come back to this topic several times below, because of its influence on contemporary mathematics education, on all levels.

What and How to Compute, and Why?

In general, because the tools are so different, the questions addressed in mathematics without computer and computational mathematics could be expected to often be different. We give below an example in the form of the *Clay Institute Millennium Prize Problem* concerning the Navier–Stokes equations. In computational mathematics the pertinent question would be: What quantity of interest of a solution to the Navier–Stokes equations can be computed to what tolerance to what cost? The prize problem formulated by a pure mathematician instead asks for a proof of existence, uniqueness and smoothness of a solution. However, there is quite a bit of evidence indicating that the Navier–Stokes equations may have turbulent solutions, which hardly can be viewed as smooth pointwise uniquely defined solutions, because turbulent solutions appear to be partly chaotic. Thus, a pure mathematician would seem to be willing to consider a question without a clear scientific relevance from applications point of view, if it offers a respectable intellectual challenge. On the other hand, in computational mathematics one may more likely focus on questions of relevance to some application. This example illustrates different attitudes in pure and computational mathematics, but of course does not give a complete picture. Some pure mathematics clearly concerns questions of scientific interest in applications, others maybe not.

The Formalists and the Constructivists

To sum up: Today there is a dividing line between pure mathematics and computational mathematics. We will below come back to this split, which originated with the birth of the computer in the 1930s in a great clash between the *formalists* (mathematics without computer) and the *constructivists* (mathematics with computer).

2.3   The Body&Soul Reform Project

Our own answers to the basic questions are presented in our mathematics education reform program, which we refer to as the *Body&Soul project* presented at www.phi.chalmers.se/bodysoul/. ← Web site

Body&Soul contains books (Applied Mathematics: Body&Soul, Vol I-III, Springer 2003, Vol IV- to appear), software and educational material and builds on a modern synthesis of Body (computation) and Soul (mathematical analysis). Body&Soul presents a synthesis of *analytical mathematics* and *computational mathematics*, where analytical mathematics is used to capture basic laws of science in mathematical notation (mathematical modeling) and to investigate qualitative aspects of such laws, and computational mathematics is needed for simulation and quantitative prediction. One way of describing analytical mathematics is to say that it is the mathematics performed with pen and paper using *symbols*, while computational mathematics is the mathematics performed using a computer. Analytical mathematics is often viewed as *pure mathematics*, although much of traditional pre-computer applied mathematics was performed using symbolic computation.

Body&Soul offers a basis for studies in science and engineering and also for further studies in mathematics, and includes modern tools of computational mathematics. The goal of Body&Soul is to present mathematics which is both understandable and useful. Body&Soul is a unique project in scope and content and is attracting quickly increasing interest.

In Appendix we reproduce the preface to the Body&Soul books. We also reproduce some sample chapters from different volumes below.

Different initiatives to reform mathematics education have been made during the last 20 years, in particular in the US under the name of Reform Calculus. Body&Soul may be viewed as a new brand of Reform Calculus with particular emphasis on a synthesis of computational and analytical mathematics. We give more substance to this aspect below.

2.4   Difficulties of Learning

Mathematics is perceived as a difficult subject by most people and feelings of insufficiency are very common, among both laymen and professionals. This is not strange. mathematics *is* difficult and demanding, just like classical music or athletics may be very difficult and demanding, which may create a lot of negative feelings for students pushed to perform. There is no way to eliminate all the difficulties met in these areas, except by trivialization. Following Einstein, one should always try to make Science and education based on Science as simple as possible, but not simpler.

In music and athletics the way out in our days is clear: the student who does not want to spend years on practicing inventions by Bach on the piano, or to become a master of high-jump, does not have to do so, but can choose some alternative activity. In mathematics this option is not available for anyone in elementary education, and not even an arts student at an American college may get away without a calculus course, not to speak of the engineering student who will have to pass several mathematics courses.

Mathematics education is thus compulsory for large groups of students, and since mathematics *is* difficult, for students on all levels, problems are bound to arise. These problems, apparent for everybody, form much of the motivation behind the task of the Mathematics Delegation.

To come to grips with this inherent difficulty of mathematics education, on all levels, we propose to offer a differentiated mathematics education, on all levels, according to the interest, ability and need of the different students. Although mathematics *is* important in our society of today, it is very possible to both survive and have a successful professional career, with very little formal training in mathematics. It is thus important to identify the real need of mathematics for different groups of students and then to shape educational programs to fill these needs.

Classical Greek or Latin formed an important part of secondary education only 50 years ago, with motivations similar to those currently used for mathematics: studies in these subjects would help develop logical thinking and problem solving skills. Today, very few students take Greek and Latin with the motivation that such studies are both difficult and of questionable usefulness for the effort invested.

One may ask if mathematics education is bound to follow the same evolution? The answer is likely to directly couple to the success of efforts to make mathematics both more understandable (less difficult) and more useful today (up to date), in contrast to the traditional education where many fail and even those who succeed may get inadequate training on inadequate topics.

Euclid's *Elementa*, with its axioms, theorems and the ruler and compasses as tools, was the canon of mathematics education for many centuries into the mid 20th century, until it quite suddenly disappeared from the curriculum along with Greek and Latin. Not because geometry ceased to be of importance, but because Euclid's geometry was replaced by *computational geometry* with the tools being Descartes analytical geometry in modern computational form.

2.5   Difficulties of Discussion

The difficulty of mathematics presents serious obstacles to discussions on mathematics and mathematics education. A pure mathematician of today would usually state that it is impossible to convey any true picture of contemporary research to anybody outside a small circle of experts. As a consequence, there is today little interaction between pure mathematics and mathematics didactics.

On the other hand, presenting essential aspects of contemporary research in computational mathematics may be possible for large audiences, including broad groups of students of mathematics. Typically, research in computational mathematics concerns the design of a computational algorithm for solving some mathematical equation, for example the Navier–Stokes equations for fluid flow. The result of the algorithm may be visualized as a movie describing some particular fluid flow, such as the flow of air around a car or airplane, and the objective of the computation could be to compute the *drag force* of a particular design, which directly couples to fuel consumption and economy. Although the details of the computation would be difficult to explain to a layman, the general structure of the computation and its meaning could be conveyed.

Thus, presenting computational mathematics to a general public appears to be possible. In particular, there is a good possibility of interaction between computational mathematics and mathematics didactics, although this connection still has to be developed.   *didactics = teaching*

We may compare with other areas of science. Many areas of science including physics, chemistry and biology share the difficulties and possibilities of mathematics. To give a deep presentation of contemporary research in these areas may be very difficult, but to convey essential aspects and its possible interest for mankind may still be possible. Of course this element is crucial when trying to raise funds for research from tax payers or private investors. To receive funding it is necessary to present reasons understandable to both politicians and the general public.

To sum up: Today there is little interaction between contemporary research in mathematics and mathematics didactics/education. Mathematics didactics needs influx from contemporary science, as does the education in any topic, and computational mathematics provides a potential source of such influx which remains to be exploited.

In fact, the sole part of mathematics education that today seems to function reasonably well is elementary arithmetics taught in elementary school (which is the basis of computational mathematics): to be able to compute with numbers is so obviously useful and the rules of computation can be made understandable even to young kids. The challenge today is to reach beyond this elementary level of plus and minus and in understandable form present modern computational mathematics with its abundance of useful applications.

2.6 Summing up the Difficulties

To sum up: Discussions on mathematics education have to struggle with some key difficulties: (i) contemporary research in pure and computational mathematics seem to live largely separated lives, and (ii) mathematics didactics largely lives a life separated from both contemporary pure and computational mathematics.

We have met both (i) and (ii) many times when trying to present our synthesis of analytical and computational mathematics. The present book represents an effort to help to overcome these difficulties in order to create a basis for a constructive discussion on mathematics education. Maybe the difficulty of presenting contemporary pure mathematics to a larger audience is insurmountable, but the situation is different for computational mathematics. The result of a computational algorithm can often be presented in a picture or movie, like for instance the fluid flow obtained from a solution of the Navier–Stokes equations. Everybody could get something out from watching such a simulation as an example of virtual reality, and could get some understanding of the principles of the computation behind the simulation.

So, it would be natural to see a lot of interaction between mathematics didactics and computational mathematics. As of now, such an interaction is still in its infancy, and we hope in particular that this little book could draw some interest from experts of mathematics didactics.

3
A Brief History of Mathematics Education

Life is good only for two things: to study mathematics and to teach it. (Poisson 1781–1840)

3.1 From Pythagoras to Calculus and Linear Algebra

The Babylonians formed a rich culture in Mesopotamia 2000-1000 BC based on advanced irrigation systems and developed mathematics for various practical purposes related to building and maintaining such systems. Using a number system with base 60, the Babylonians could do arithmetics including addition, subtraction, multiplication and division and could also solve quadratic equations. The development continued in ancient Greece 500 BC-100 AD, where the schools of Pythagoras and Euclid created eternal foundations of *arithmetics* and *geometry*. The next leap came with the *calculus* of Leibniz and Newton preceded by the *analytic geometry* by Descartes, which opened for the *scientific revolution* starting in the 18th century and leading into our time. calculus (or differential and integral calculus) is the mathematical theory of functions, derivatives, integrals and differential equations.

Calculus (also referred to as mathematical analysis) developed over the centuries with important contributions from many great mathematicians such as Cauchy and Weierstrass and found its form in the present curriculum of science and engineering education in the beginning of the 20th century. Together with *linear algebra* including vector and matrix calculus introduced in the 1950s, calculus today forms the core of mathematics education at the university level, and simplified forms thereof fill the mathematics curricula in secondary schools. The introduction of linear algebra was probably stimulated by the development of the computer, but both calculus and linear algebra are still presented as if the computer does not exist. The foreword of a new standard calculus text usually pays tribute to an idea that calculus found its form in the early 20th century and that a new book can only polish on a forever given picture.

The classical *curriculum* of the university appearing in the 15th century was the *trivium* (intersection of three roads) containing *grammar, rethorics* and *dialectics*. There was also a *quadrivium* containing *arithmetics, geometry, music* and *astronomy,* all viewed as parts of mathematics following the tradition of the Pythagorean school. Arithmetics and geometry connected to *thought* and music and astronomy to *experience* and reflect distinctions between mathematics and the sciences still valid in our time.

rhetoric [margin annotation]

A further distinction was made between arithmetics and music being *discrete* (or *digital*), while Geometry and Astronomy were thought of as *continuous*. We will return to this aspect below, which with modern aspects of *digital computation* gets new light.

3.2   From von Neumann into Modern Society

Now an established myth. [margin annotation]

We all know that the modern computer is fundamentally changing our society. One may describe this development alternatively as a *mathematics revolution* reflecting that the computer may be thought of as a *mathematical machine*. More precisely, the modern computer is referred as the *von Neumann Machine* after the famous mathematician John von Neumann who first formulated the mathematical principles of modern computer design and computer programming in the early 1940s.

Von Neumann followed up on a long tradition within mathematics of constructing machines for automated computation, from the mechanical calculators of Pascal and Leibniz doing elementary arithmetics, over Babbage's *Difference Engine* and *Analytical Computing Engine* designed to solve differential equations, to the theoretical *Turing Machine* capable of mimicing the action of any conceivable *computer*. The motivation of all these machines was to enable *automated computation* for various purposes, typically connected to calculus.

We may view our modern *information society* as the *society of automated computation*. Similarly, we may view the modern *industrial society* as the *society of automated production*. Altogether, we may view our *modern society* as a combination of the industrial society and the information society.

In the chapter *What is Mathematics?* below, we give examples of the use of mathematics in our information society, with automated computation on words, images, movies, sound, music which form the essence of the new world of *virtual reality*.

3.3 Mathematics Education and the Computer

We observed above that mathematics education in its current canonical form of calculus, is basically the same as before the computer revolution. We argue that this state of affairs is not motivated from either scientific or applications point of view. We believe that the computer is radically changing calculus and that this change has to be seen in the mathematics curriculum.

This point of view is not the standard one giving the rationale of the present mathematics curriculum, which is much more modest, something like: ok, the computer is now here, but the essence of calculus remains the same and the education should not be radically changed in any respect.

This book is about this issue: does the computer create a new form of calculus to be presented in education, a *reform calculus*? We answer: yes! From this basis we present a program for such a new form of calculus in the form of *Computational Mathematical Modeling* including computation as a new element.

Proponents of standard calculus would answer: no! (or: hardly!). The standard curriculum should last for another 100 years. Nothing much has to be changed.

So there we stand today, without any clear picture emerging. Standard calculus still dominates the scene, but reform calculus is quickly growing. And the change could come quickly. It is possible to change from standard to reform calculus instantly. We did this for a group of chemistry students at Chalmers, and we are just waiting to see other students follow...

3.4 The Multiplication Table

The multiplication table may be viewed as the corner stone of elementary mathematics education. The educated man of the 17th century did not necessarily master the multiplication table, as this knowledge was something for practitioners like merchants and carpenters, not for men of learning. This is illustrated in the famous diary by Samuel Pepys 1660–69 during his studies at the University of Cambridge, where he describes his difficulties of learning to master the table: "the most difficult subject he had ever encountered". This was the same Pepys that created the modern English fleet and became the chairman of the Royal Society and publisher of Newton's monumental *Principia Mathematica.*

Despite Pepys' difficulties, the multiplication table and spelling formed the very core of the public school as it was formed in the mid 19th century. These topics gave instruments for exercising control and selection in the school system, as they gave objective criteria for sorting students to serve in the industrial society.

3.5   Again: What?

In the information society of today, difficulties of spelling can be compensated by using a *word processor* with a *spell checker*, and would not necessarily be a stumbling block for a career in academics, administration or politics for all the talented and intelligent people with some dyslectic syndrome. Likewise, today the pocket calculator can easily compensate a lack of mastery of the multiplication table, not to speak of the even more complicated algorithm for long division. Thus, today the corner stones of traditional education, spelling and the multiplication table, seem to be loosing importance as pillars of elementary education.

So if the multiplication table and long division no longer serve as the canon of elementary mathematics education, what could/should then be taught? What could/should be the purpose of elementary mathematics education? Or should the old canon be resurrected?

4
What is Mathematics?

> The question of the ultimate foundations and the ultimate meaning of mathematics remains open; we do not know in what direction it will find its final solution or whether a final objective answer may be expected at all. "Mathematizing" may well be a creative activity of man, like language or music, of primary originality, whose historical decisions defy complete objective rationalization. (Weyl)

> The universal mathematics is, so to speak, the logic of the imagination. (Leibniz)

4.1 Introduction

We start out by giving a very brief idea of the nature of mathematics and the role of mathematics in our society. This is the first chapter of Body&Soul Vol I.

4.2 The Modern World: Automated Production and Computation

The mass consumption of the *industrial society* is made possible by the *automated mass production* of material goods such as food, clothes, housing, TV-sets, CD-players and cars. If these items had to be produced by hand, they would be the privileges of only a select few.

FIGURE 4.1. First picture of book printing technique (from Danse Macabre, Lyon 1499)

Analogously, the emerging *information society* is based on mass consumption of *automated computation* by computers that is creating a new "virtual reality " and is revolutionizing technology, communication, administration, economy, medicine, and the entertainment industry. The information society offers immaterial goods in the form of knowledge, information, fiction, movies, music, games and means of communication. The modern PC or lap-top is a powerful computing device for mass production/consumption of information e.g. in the form of words, images, movies and music.

Key steps in the automation or mechanization of production were: Gutenberg's book printing technique (Germany, 1450), Christoffer Polhem's automatic machine for clock gears (Sweden, 1700), The Spinning Jenny (England, 1764), Jacquard's punched card controlled weaving loom (France, 1801), Ford's production line (USA, 1913), see Fig. 4.1, Fig. 4.2, and Fig. 4.3.

Key steps in the automation of computation were: Abacus (Ancient Greece, Roman Empire), Slide Rule (England, 1620), Pascal's Mechanical Calculator (France, 1650), Babbage's Difference Machine (England, 1830), Scheutz' Difference Machine (Sweden, 1850), ENIAC Electronic Numerical Integrator and Computer (USA, 1945), and the Personal Computer PC (USA, 1980), see Fig. 4.5, Fig. 4.6, Fig. 4.7 and Fig. 4.8. The Difference Machines could solve simple differential equations and were used to compute tables of elementary functions such as the logarithm. ENIAC was one of the first modern computers (electronic and programmable), consisted of 18,000 vacuum tubes filling a room of 50×100 square feet with a weight of 30 tons and energy consuming of 200 kilowatts, and was used to solve

FIGURE 4.2. Christoffer Polhem's machine for clock gears (1700) and the Spinning Jenny (1764).

FIGURE 4.3. Ford assembly line (1913).

the differential equations of ballistic firing tables as an important part of the Allied World War II effort. A modern laptop at a cost of $2000 with a processor speed of 2 GHz and internal memory of 512 Mb has the computational power of hundreds of thousands of ENIACs.

Automation is based on frequent repetition of a certain *algorithm* or scheme with new data at each repetition. The algorithm may consist of a sequence of relatively simple steps together creating a more complicated process. In automated manufacturing, as in the production line of a car factory, physical material is modified following a strict repetitive scheme, and in automated computation, the 1s and 0s of the microprocessor are modified billions of times each second following the computer program. Similarly, a *genetic code* of an organism may be seen as an algorithm that generates a living organism when realized in interplay with the environment. Realizing a genetic code many times (with small variations) generates populations of organisms. Mass-production is the key to increased complexity following the patterns of nature: elementary particle $\rightarrow$ atom $\rightarrow$ molecule and molecule $\rightarrow$ cell $\rightarrow$ organism $\rightarrow$ population, or the patterns of our society: individual $\rightarrow$ group $\rightarrow$ society or computer $\rightarrow$ computer network $\rightarrow$ global net.

In this analogy the environment is the input, the organism is the output. This is a very rough analogy.

FIGURE 4.4. Computing device of the Inca Culture.

4.3   The Role of Mathematics

Mathematics may be viewed as the language of computation and thus lies at the heart of the modern information society. Mathematics is also the language of science and thus lies at the heart of the industrial society that grew out of the *scientific revolution* in the 17th century that began when Leibniz and Newton created *calculus*. Using calculus, basic laws of mechanics and physics, such as Newton's law, could be formulated as *mathematical models* in the form of *differential equations*. Using the models, real phenomena could be *simulated* and controlled (more or less) and industrial processes could be created.

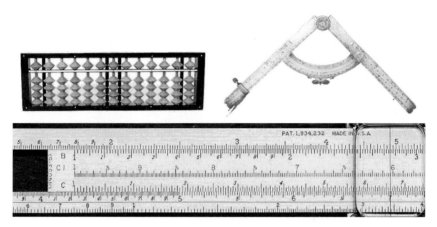

FIGURE 4.5. Classical computational tools: the Abacus, Galileo's Compass and the Slide Rule.

The mass consumption of both material and immaterial goods, considered to be a corner-stone of our modern democratic society, is made possible through automation of production and computation. Therefore, mathematics forms a fundamental part of the technical basis of the modern society revolving around automated production of material goods and automated computation of information.

The vision of virtual reality based on automated computation was formulated by Leibniz already in the 17th century and was developed further by Babbage with his Analytical Engine in the 1830s. This vision is finally being realized in the modern computer age in a synthesis of Body & Soul of mathematics.

We now give some examples of the use of mathematics today that are connected to different forms of automated computation.

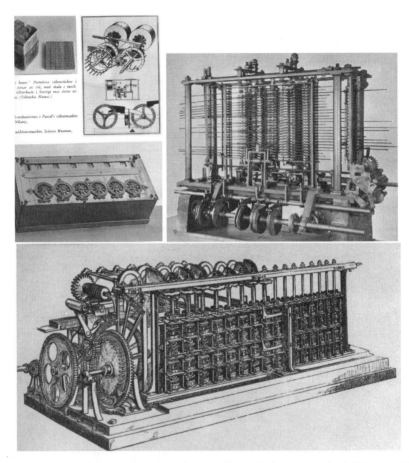

FIGURE 4.6. Napier's Bones (1617), Pascal's Calculator (1630), Babbage's Difference Machine (1830) and Scheutz' Swedish Difference Machine (1850).

FIGURE 4.7. Odhner's mechanical calculator made in Göteborg, Sweden, 1919–1950

FIGURE 4.8. ENIAC Electronic Numerical Integrator and Calculator (1945).

4.4   Design and Production of Cars

In the car industry, a model of a component or complete car can be made using Computer Aided Design CAD. The CAD-model describes the geometry of the car through mathematical expressions and the model can be displayed on the computer screen. The performance of the component can then be tested in computer simulations, where differential equations are solved through massive computation, and the CAD-model is used as input of geometrical data. Further, the CAD data can be used in automated production. The new technique is revolutionizing the whole industrial process from design to production.

4.5   Navigation: From Stars to GPS

A primary force behind the development of geometry and mathematics since the Babylonians has been the need to navigate using information from the positions of the planets, stars, the moon and the sun. With a clock and a sextant and mathematical tables, the sea-farer of the 18th century could determine his position more or less accurately. But the results depended strongly on the precision of clocks and observations and it was easy for large errors to creep in. Historically, navigation has not been an easy job.

During the last decade, the classical methods of navigation have been replaced by GPS, the Global Positioning System. With a GPS navigator in hand, which we can buy for a couple of hundred dollars, we get our coordinates (latitude and longitude) with a precision of 50 meters at the press of a button. GPS is based on a simple mathematical principle known already to the Greeks: if we know our distance to three points is space with known coordinates then we can compute our position. The GPS uses this principle by measuring its distance to three satellites with known positions, and then computes its own coordinates. To use this technique, we need to deploy satellites, keep track of them in space and time, and measure relevant distances, which became possible only in the last decades. Of course, computers are used to keep track of the satellites, and the microprocessor of a hand-held GPS measures distances and computes the current coordinates.

The GPS has opened the door to mass consumption in navigation, which was before the privilege of only a few.

4.6   Medical Tomography

The computer tomograph creates a picture of the inside of a human body by solving a certain integral equation by massive computation, with data coming from measuring the attenuation of very weak x-rays sent through

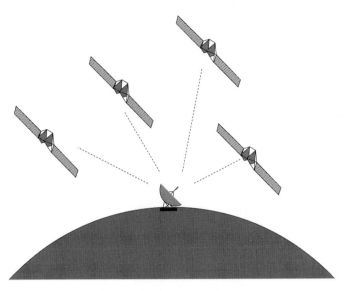

FIGURE 4.9. GPS-system with 4 satellites.

the body from different directions. This technique offers mass consumption of medical imaging, which is radically changing medical research and practice.

4.7 Molecular Dynamics and Medical Drug Design

The classic way in which new drugs are discovered is an expensive and time-consuming process. First, a physical search is conducted for new organic chemical compounds, for example among the rain forests in South America. Once a new organic molecule is discovered, drug and chemical companies license the molecule for use in a broad laboratory investigation to see if the compound is useful. This search is conducted by expert organic chemists who build up a vast experience with how compounds can interact and which kind of interactions are likely to prove useful for the purpose of controlling a disease or fixing a physical condition. Such experience is needed to reduce the number of laboratory trials that are conducted, otherwise the vast range of possibilities is overwhelming.

The use of computers in the search for new drugs is rapidly increasing. One use is to makeup new compounds so as to reduce the need to make expensive searches in exotic locations like southern rain forests. As part of this search, the computer can also help classify possible configurations of molecules and provide likely ranges of interactions, thus greatly reducing the amount of laboratory testing time that is needed.

License is necessary so you can sell & profit, but not sufficient.

There are also clinical trials required.

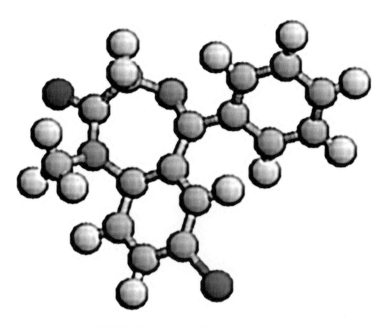

FIGURE 4.10. The Valium molecule.

4.8   Weather Prediction and Global Warming

Weather predictions are based on solving differential equations that describe the evolution of the atmosphere using a super computer. Reasonably
reliable predictions of daily weather are routinely done for periods of a few
days. For longer periods. the reliability of the simulation decreases rapidly,
and with present day computers daily weather predictions for a period of
two weeks are impossible.

However, forecasts over months of averages of temperature and rainfall
are possible with present day computer power and are routinely performed.

Long-time simulations over periods of 20-50 years of yearly temperature-
averages are done today to predict a possible *global warming* due to the use
of fossil energy. The reliability of these simulations are debated.

4.9   Economy: Stocks and Options

The Black-Scholes model for pricing options has created a new market of
so called derivative trading as a complement to the stock market. To cor-
rectly price options is a mathematically complicated and computationally
intensive task, and a stock broker with first class software for this purpose
(which responds in a few seconds), has a clear trading advantage.

4.10   The World of Digital Image, Word and Sound

The modern computer and internet now offer us a flood of information, science or entertainment in digital form. Through *Data Mining* we can search massive *Data Bases* for *information*. By *Data Compression* of images, words and sound we can store and transmit huge Data Bases. This whole new world of our information society is based on mathematical algorithms for compressing and searching data.

4.11   Languages

Mathematics is a *language*. There are many different languages. Our mother tongue, whatever it happens to be, English, Swedish, or Greek, is our most important language, which a child masters quite well at the age of three. To learn to write in our native language takes longer time and more effort and occupies a large part of the early school years. To learn to speak and write a foreign language is an important part of secondary education.

Language is used for *communication* with other people for purposes of cooperation, exchange of ideas or control. Communication is becoming increasingly important in our society as the modern means of communication develop. Communication promotes communication.

Using a language we may create *models* of phenomena of interest, and by using models, phenomena may be studied for purposes of *understanding* or *prediction*. Models may be used for *analysis* focussed on a close examination of individual parts of the model and for *synthesis* aimed at understanding the interplay of the parts that is understanding the model as a whole. A *novel* is like a model of the real world expressed in a written language like English. In a novel the characters of people in the novel may be analyzed and the interaction between people may be displayed and studied.

The ants in a group of ants or bees in a bees hive also have a language for communication. In fact in modern biology, the interaction between cells or proteins in a cell is often described in terms of entities "talking to each other".

It appears that we as human beings use our language when we *think*. We then seem to use the language as a model in our head, where we try various possibilities in *simulations* of the real world: "If that happens, then I'll do this, and if instead that happens, then I will do so and so...". Planning our day and setting up our calendar is also some type of modeling or simulation of events to come. Simulations by using our language thus seems to go on in our heads all the time.

There are also other languages like the language of musical notation with its notes, bars, and scores. A musical score is like a model of the real music. For a trained composer, the model of the written score can be very close to the real music. For amateurs, the musical score may say very little, because the score is like a foreign language which is not understood.

4.12   Mathematics as the Language of Science

Mathematics has been described as the language of science and technology including mechanics, astronomy, physics, chemistry, and topics like fluid mechanics, solid mechanics, and electromagnetics. The language of mathematics is used to deal with *geometrical* concepts like *position* and *form* and *mechanical* concepts like *velocity, force* and *field.* More generally, mathematics serves as a language in any area that includes *quantitative* aspects described in terms of *numbers*, such as economy, accounting, and statistics. Mathematics serves as the basis for the modern means of electronic *communication* where information is coded as sequences of 0's and 1's and is transferred, manipulated, and stored. *0's & 1's with boundaries with no boundaries there is no structure.*

The words of the language of mathematics often are taken from our usual language, like *points, lines, circles, velocity, functions, relations, transformations, sequences, equality*, and *inequality.*

A mathematical word, term or concept is supposed to have a specific meaning defined using other words and concepts that are already defined. This is the same principle as is used in a thesaurus, where relatively complicated words are described in terms of simpler words. To start the definition process, certain fundamental concepts or words are used, which cannot be defined in terms of already defined concepts. Basic relations between the fundamental concepts may be described in certain *axioms*. Fundamental concepts of Euclidean geometry are *point* and *line*, and a basic Euclidean axiom states that through each pair of distinct points there is a unique line passing. A *theorem* is a statement derived from the axioms or other theorems by using logical reasoning following certain rules of logic. The derivation is called a *proof* of the theorem.

4.13   The Basic Areas of Mathematics

The basic areas of mathematics are

- geometry *Is that it?*

- algebra

- analysis.

Geometry concerns objects like *lines, triangles, circles*. algebra and analysis is based on *numbers* and *functions*. The basic areas of mathematics education in engineering or science education are

- calculus

- linear algebra.

Calculus is a branch of analysis and concerns properties of functions such as *continuity*, and operations on functions such as *differentiation* and *integration*. Calculus connects to linear algebra in the study of *linear functions* or linear transformations and to *analytical geometry*, which describes geometry in terms of numbers. The basic concepts of calculus are

- function

- derivative

- integral.

Linear algebra combines geometry and algebra and connects to analytical geometry. The basic concepts of linear algebra are

- vector

- vector space

- projection, orthogonality

- linear transformation.

The Body&Soul series of books teach the basics of calculus and linear algebra, which are the areas of mathematics underlying most applications.

4.14   What is Science?

The theoretical kernel of *natural science* may be viewed as having two components

- formulating equations (modeling),

- solving equations (computation).

Together, these form the essence of *mathematical modeling* and *computational mathematical modeling*. The first really great triumph of science and mathematical modeling is Newton's model of our planetary system as a set of differential equations expressing Newton's law connecting force, through the inverse square law, and acceleration. An *algorithm* may be seen as a strategy or constructive method to solve a given equation via computation. By applying the algorithm and computing, it is possible to simulate real phenomena and make predictions.

Traditional techniques of computing were based on symbolic or numerical computation with pen and paper, tables, slide ruler and mechanical calculator. Automated computation with computers is now opening new possibilities of simulation of real phenomena according to Natures own principle of massive repetition of simple operations, and the areas of applications are quickly growing in science, technology, medicine and economics.

Mathematics is basic for both steps (i) formulating and (ii) solving equation. Mathematics is used as a language to formulate equations and as a set of tools to solve equations. *You want to be famous?*

Fame in science can be reached by formulating or solving equations. The success is usually manifested by connecting the name of the inventor to the equation or solution method. Examples are legion: Newton's method, Euler's equations, Lagrange's equations, Poisson's equation, Laplace's equation, Navier's equation, Navier–Stokes' equations, Boussinesq's equation, Einstein's equation, Schrödinger's equation, Black-Scholes formula..., some of which we will meet below.

4.15   Mathematics is Difficult: Choose Your Own Level of Ambition

First, we have to admit that mathematics is a difficult subject, and we see no way around this fact. Secondly, one should realize that it is perfectly possible to live a happy life with a career in both academics and industry with only elementary knowledge of mathematics. There are many examples including Nobel Prize winners. This means that it is advisable to set a level of ambition in mathematics studies which is realistic and fits the interest profile of the individual student. Many students of engineering have other prime interests than mathematics, but there are also students who really like mathematics and theoretical engineering subjects using mathematics. The span of mathematical interest may thus be expected to be quite wide in a group of students following a course based on the Body&Soul series of books, and it seems reasonable that this would be reflected in the choice of level of ambition.

4.16   Some Parts of Mathematics are Easy

On the other hand, there are many aspects of mathematics which are not so difficult, or even "simple", once they have been properly understood. Thus, Body&Soul Vol I-III contains both difficult and simple material, and the first impression from the student may give overwhelming weight to the former. To help out we have in Body&Soul collected the most essential nontrivial facts in short summaries in the form of *Calculus Tool Bag I and II, Linear Algebra Tool Bag, Differential Equations Tool Bag, Applications Tool Bag, Fourier Analysis Tool Bag* and *Analytic Functions Tool Bag*. The reader will find the tool bags surprisingly short: just a couple pages, altogether say 15-20 pages. If properly understood, this material carries a long way and is "all" one needs to remember from the math studies for further studies and professional activities in other areas. Since the first three volumes of the Body&Soul series of books contains about 1200 pages it means 50-100 pages of book text for each one page of summary. This means that the books give more than the absolute minimum of information and has the ambition to give the mathematical concepts a perspective concerning both history and applicability today. So we hope the student does not get turned off by the quite a massive number of words, by remembering that after all 15-20 pages captures the essential facts. During a period of say one year and a half of math studies, this effectively means about one third of a page each week!

Summary

4.17 Increased/Decreased Importance of Mathematics

Body&Soul reflects both the increased importance of mathematics in the information society of today, and the decreased importance of much of the analytical mathematics filling the traditional curriculum. The student should thus be happy to know that many of the traditional formulas are no longer such a must, and that a proper understanding of relatively few basic mathematical facts can help a lot in coping with modern life and science.

5
Virtual Reality and the Matrix

All thought is a kind of computation. (Hobbes)

5.1 Virtual Reality

The world of *virtual reality* is created using computational mathematics. Virtual reality includes the flood of computer games with increasing realism, but also e.g. the rapidly developing field of *medical imaging*, allowing a virtual patient or organ to be created, which the surgeon can use to plan and perform the actual operation on the real patient. Virtual patients can also be used in simulators for the training of surgeons, instead of real bodies. Medical imaging is based on computational mathematics where information from very weak x-rays or acoustic/electro-magnetic waves penetrating the body is used in a massive computation on a computer to create a 3D image from data measured outside the body.

Virtual reality seems unlimited in scope, with just the computational power setting the limits. To create virtual reality we need tools of imaging and tools of simulation to give the created images life, and the tools come from computational mathematics.

5.2 Digital Cameras

→ It could also be hexagonal.

Many of us now use digital cameras, and are familiar with the representation of a 2D picture as a rectangular array of pixels with each pixel being a small square with a certain color. For example a picture of say 2000×1000 pixels would be a 2 megapixel picture, which would occupy roughly 1 Mb on the memory card of the camera. We also know that we may store the picture in compressed form using less memory (e.g the JPEG-format), and that we may subject the picture to various transformations all based on different algorithms from computational mathematics.

5.3 MP3

In 1987 the German Fraunhofer Institute started to develop a new compressed form of digital representation of sound (audio coding), which resulted in the MP3 (ISO-MPEG Audio Layer-3) standard, which is now used extensively. A full digital representation of a sound signal of a bandwidth of 44 kHz would require 1.4 Mbits/second, while MP3 typically needs only 112 kbps with a reduction factor of 12, without significant loss of quality. MP3 acts like a mathematical filter simplifying the sound signal without changing the impression by the ear.

5.4 Matrix and Marilyn Monroe

Ugh!

The film *Matrix* has become a cult movie, not without good reason. Matrix connects to basic aspects of human spiritual life concerning our perceptions of what *reality* may be. We all know that what we directly perceive as reality is filtered through our 5 senses and that there are many aspects of light, sound, smell, taste, and tactile sensations "out there" that we miss. And we may also feel that there may be even more than that. Different religions seek to present "virtual realities", often believed to be more "true" than our every-day supposedly distorted images.

Young people now spend several hours a day in the virtual reality of the computer. We may as adults view this as silly and meaningless, but the kids may have a different perception. Books also offer a kind of virtual reality, like all the endless stories told during the many years homo sapiens spent in the caves before turning on the lap top computer in the high-rise flat or internet cafe, and thus we seem to have both a talent and a need of virtual reality, in different forms. What makes a good book better than a good computer game or interactive computer novel, where we can directly enter the world of the imagined characters and interact with them? Wouldn't it be nice to be able to interact with real persons like Marilyn Monroe or

Einstein in a virtual reality, rather than just looking at some photos in a book? Or interact with fictitious people from the great novels, like Captain Nemo, Madame Bovary, Charles Swann, Molly Bloom, and others?

I like the way the boots are drawn.

FIGURE 5.1. Captain Nemo taking a star sight from the deck of the Nautilus.

During the day? Or at night?

Captain Nemo's boots: Consider them as an example of surface structure.

6
Mountain Climbing

Every new body of discovery is mathematical in form, because there is no other guidance we can have. (Darwin)

The *key principle* of our whole educational system states that *education is based on science*, more precisely *education of today should be based on contemporary science*. We touched this principle above and will return to it below in some more detail.

We now present an aspect of *mountain climbing* illustrating the key principle or the role of science in education, or the role of a scientist as a discoverer of scientific truths.

In mountain climbing one important technique is to facilitate climbing by attaching a rope to bolts driven into glitches in the rock. Once the rope is fastened, it may serve as a safety measure for many climbers, and thus facilitate climbing of difficult passages to reach new heights. However, the first climber will have to climb to a new level without a rope and therefore the role and task of the first climber is very crucial. Snelson is a first climber.

Now, in education as a kind of climbing process to higher levels of knowledge and understanding, the scientist has the role of first climbing to a new level to secure a fixed point in a Mountain of Knowledge. This may be a risky and difficult task, but if successful it may help many other scientists and students to follow and reach that new level.

The analogy also illustrates the role of the most novel scientific discovery in setting the position of all the bolts and the connecting rope below the new level. Since the entire rope has to hang together, it may be necessary to change the position of many old bolts to fit the position of the just one new bolt

bolt on a higher level. This reflects a principle of unity in science, which does not mean that there may not be several ropes leading to a mountain top, but that each rope has to be connected.

In particular, we want this way to emphasize the fact that a new scientific discovery may influence also the education on all levels from elementary and up. A new scientific discovery about the genetic code or the nature of gravitation may directly influence education on all levels. We see this happening in biology and physics, and in social and political sciences along with the rapid changes of society, and of course in different areas of engineering along with advancements in technology, and could expect that it should also happen in mathematics: a new bolt on a new level may change the entire rope. *A bolt = a point of attachment.*

In our own teaching the effect may be as follows: in a course for *second year* engineering students, we *start* by posing the top problem of contemporary science of *turbulence*, and during the course we work our way from basic elements towards a solution of this problem. How this is done is explained in the chapter *Turbulence and the Clay Prize* below. We thus put contemporary research into basic mathematics education, as is necessary to do in for example *molecular biology*, where even a 10 year old knowledge may have passed a best-before date. *may have become obsolete.*

FIGURE 6.1. "A born climber's appetite for climbing is hard to satisfy; when it comes upon him he is like a starving man with a feast before him; he may have other business on hand, but it must wait." (Mark Twain)

7
Scientific Revolutions

I know that the great Hilbert said "We will not be driven out from the paradise Cantor has created for us", and I reply "I see no reason to walking in". (R. Hamming)
for

7.1 Galileo and the Market

During the evolution of our culture, the shaman of the primitive religion was replaced by the priest, who is now being replaced by the scientist, as the chief connection of ordinary people to the Creator. In 1633 the scientist Galileo under a threat of death penalty from the Catholic Church had to deny his conviction that the planets of our solar system including the earth orbit around the sun. That time the church seemed to be the winner, but the ideas of Galileo initiated the scientific revolution which changed the game completely: the important disputes no longer occur between the church and science but rather between different disciplines and schools of science. Today, investors on the Market are hoping to see new scientific discoveries that can be patented and can generate fortunes. So the fight to gain ideological or economical influence in science has a long history and continues today under new rules and conditions of competition.

7.2   Thomas Kuhn and Scientific Revolutions

According to Thomas Kuhn, the author of the famous book "The Structure of Scientific Revolutions", science develops through a series of scientific revolutions, where new paradigms or sets of basic principles or beliefs replace old ones. A *paradigm* includes (i) formulation of questions, (ii) selection of methods to answer the questions and (iii) definition of areas of relevance. A new paradigm develops when certain questions cannot find answers with available theories/methods, and therefore new theories/methods are developed which give new answers of new relevance, but also pose new questions, and so on.

The scientific revolution of the 17th century was Newton's mechanics including his theory of gravitation giving Galileo convincing mathematical support.

Kuhn carefully analyzed the scientific revolution of the new physics of Quantum Mechanics emerging in the 1920s replacing/extending Newtons Mechanics. Mathematically, this corresponds to replacing Newton's equations by Schrödinger's. Physically, it corresponds to viewing e.g. an electron as a wave rather than a particle, with solutions of Schrödinger's equation being referred to as "wave functions". It also corresponds to a shift towards a partly probabilistic view instead of the fully deterministic view of classical mechanics, expressed through Heisenberg's Uncertainty Relation.

Kuhn noticed that a scientific revolution could be perceived as "invisible" in the sense that many scientists would not be aware of (or simply deny) an ongoing shift of paradigm, in a (subconscious) reaction to save the old system.

So shifts of paradigms in science do occur, and obviously are of crucial importance in the evolution of science. They correspond to key steps in the evolution of life on Earth, like the revolutionary new concept of the first mammals developed during the time of the Dinosaurs 65 millions years ago, which quickly took over the scene (with some help from volcanos or meteors supposedly ending the era of the Dinosaurs).

7.3   Shift of Paradigm in Mathematics?

Today, we may witness such a shift of paradigm from mathematics without computer to mathematics with computer, although according to Kuhn it could be expected to be "invisible" to many actors. As in all shifts of paradigm, the situation may seem very confused to the outside observer, with the old and the new viewpoints both trying to get the attention. And of course one may expect the fight to partly be tough because important values and investments are at stake. So we invite the reader to make observations and draw conclusions, and to be careful not to accept anything for given unless good reasons to do so are presented.

In physics we see today a dispute between the new physics of string theory and quantum mechanics, and nobody seems today to be able to predict the outcome. String theory may change physics just as quantum mechanics did, or it may disappear as one of the unhappy mutations in the evolution of science. Time will tell...

The prospects for computational mathematics closely connect to the prospects for the whole IT-sector. In the short run it may be difficult to tell what to invest in, as a market analyst probably would say, but in the long run the prospects are formidable.

7.4   Quarrelling Mathematicians

It is important to notice that also mathematics as a science has met serious controversies. The fight during the 1930s between the *formalist* and *constructivist* schools within mathematics is described below in the chapter *Do Mathematicians Quarrel?*, reprinted from Body&Soul Vol I.

A very short summary goes as follows: The leader of the formalist school was the mathematician Hilbert (who formulated the famous 23 Problems at the World Mathematics Congress in Paris in 1900). His hope was to give a rigorous basis to mathematics based on "finitary" principles. The hope of Hilbert was refuted first by Gödel, who showed that there are mathematical truths, which can not be proved "finitary", and second by Turing (the inventor of the principle of the computer), who proved that there are numbers which are uncomputable by "finitary" methods. Hilbert's ideas were simply incompatible with those of Gödel and Turing, but the controversy was never resolved. Instead, mathematics split into pure mathematics disregarding the limitations given by Gödel and Turing, and computational mathematics working with the possibilities and limitations of "finitary" or computational methods. Pure mathematics took charge of the mathematics departments and computational mathematics developed outside.

Today the rapidly increasing access to cheap computational power brings new life to these questions, and this time maybe a healthy synthesis may be created, which is basic idea behind Body&Soul.

7.5   Change of Paradigm? or Not?

A crucial question for mathematics education today is thus if: (B1) there is a crisis in mathematics education today, and (B2) a change of paradigm or scientific revolution caused by the computer is now going on in mathematics as a science.

There are many experts and laymen who claim that B1 and B2 are true, and there are many who claim the opposite. This is normal in a change of paradigm: some see it coming before others. When everybody sees the same thing, the change of paradigm is already over. There are many examples of such reactions from the last decades of the information society with the changes of paradigm caused by word processors, mobile telephones and internet, and of course many throughout the history of the industrial society.

7.6   To Prove or Prove Not So

To prove that Saddam Hussein had no weapons of mass-destruction (if he didn't), was probably more difficult than proving that he actually had (if he did). So to be really sure that a change of paradigm *is not* going on, may be more difficult (because you have to go through *all* evidence), than to be really sure that it *is* (because *one* piece of evidence may be enough). One single site for the production of weapons of mass destruction in operation would have been enough. But no single one was found.

7.7   The Role of Text Books

According to Thomas Kuhn a scientific paradigm is described in its text books. The writers of text books interpret the paradigm and transmit it to new generations of students. The large number of standard Calculus books, all quite similar, represent a traditional paradigm formed in the beginning of the 20th century. The Body&Soul reform project with its sequence of text books represents our own effort to present a new paradigm including computational mathematics. Time will show if our efforts had any effect or not. At least we tried.

8
Education is Based on Science

> In science one tries to tell people, in such a way as to be understood
> by everyone, something that no one ever knew before. But in poetry,
> it's the exact opposite. (Dirac 1902-1984)

The education at a modern university is assumed to have a scientific
basis. This means that the material presented should reflect the *current*
standpoint of science, not the standpoint say 100 years ago. This is very
evident in scientific disciplines such as physics, chemistry and biology, where
it would be impossible to neglect the discovery of the electron (Nobel Prize
1906) or the molecular structure of DNA (Nobel Prize 1962).

The education programs in primary and secondary schools necessarily
have to be simplified versions of the corresponding university programs,
and thus also has to have a scientific basis. Even in pre-school education,
it would today be impossible to deny the existence of bacteria (discovered
in the late 19th century), and the electron would necessarily appear in
secondary school physics.

8.1 The Scientific Basis of Standard Calculus

In mathematics education however, the situation seems to be different. The
standard calculus book of today is modeled on a pattern set during the 19th
century based on the work by e.g. Cauchy and Weierstrass, and the impact
of modern computational methods is very small. Authors of calculus books
(and there are many) seem to assume that calculus in this classical form

will last forever, and each new calculus book seems to be a rewrite of an already existing calculus book. In particular, students of calculus would find no reason to believe that the scientific basis of calculus has changed because of the development of the computer and computational methods during the 20th century, just as if in physics the electron had not been discovered.

The standard view would thus be that nothing fundamental in calculus has changed from the end of the 19th to the end of the 20th century: derivatives, integrals and differential equations are eternal objects and their meaning and role in science do not depend on the existence of the computer. With this view the scientific basis of calculus has not changed during the last 100 years in any essential way and therefore the calculus education does not have to be changed either.

8.2   A New Scientific Basis

Our point is different from the standard view: we firmly believe that the computer has fundamentally changed calculus as a science. We share this view with many scientists and mathematicians, but the standard view is still dominating. There are things that have not changed, but there are also aspects where the perspective indeed has changed a lot. Thus, the questions and answers partly are very different with the computer than without. Our Body&Soul reform program is based on this conviction. An example: with the computer one can computationally solve mathematical equations such as the Navier–Stokes equations for fluid flow, for which analytical solution has been and most likely always will be impossible. The impact of computational methods in fluid dynamics as a science and area of engineering practice is already strong and will grow stronger. For the first time in human history it is possible to computationally simulate *turbulent* fluid flow and uncover its mysteries.

We return to this topic below in a discussion on the famous *Clay Prize* offering $1 million for an analytical mathematical proof of existence and uniqueness of solutions to the Navier–Stokes equations.

Many mathematicians would today claim that our standpoint is not necessarily the right one. They would advocate that it is possible to present calculus as a teacher or author without any deep insight into modern computational mathematics. We argue that such a deep insight indeed is necessary. We meet this problem constantly when we seek to present our reform calculus: without a common scientific basis communication becomes very difficult. We already addressed this issue above and will return to this key aspect below.

8.3   The Scientific Basis of Body&Soul

The Body&Soul reform program is an off-spring of our scientific work within computational mathematics over the last 30 years focused on developing a general methodology for computational solution of differential equations. We thus believe that Body&Soul has a scientific basis which reflects a current stand-point of science.

The fact that Body&Soul is based on contemporary science represents one of the difficulties of discussion already pointed to. It appears to us impossible to have any sensible reaction to Body&Soul without a fairly deep knowledge of contemporary research in computational mathematics.

Our main message is that indeed computation gives calculus a new meaning and role in science and education as well as a new scientific basis.

Our experience is that a serious discussion of calculus reform without a common scientific basis including modern computation is very difficult, and this difficulty will remain as long as computational methods are not considered to be an integral part of calculus.

To sum up: Our reform calculus program is motivated by our scientific work on computational methods. It is our conviction based on our experience from this work that computation is now giving calculus a new meaning and role as a science, and that the education and use of calculus in applications will have to be reformed to properly take the possibilities opened into account.

We believe that the impact of computation on mathematics can no longer be neglected.

9

The Unreasonable Effectiveness of Mathematics in the Natural Sciences?

I don't like it (quantum mechanics), and I'm sorry I ever had anything to do with it. (Schrödinger)

In 1960, E.P. Wigner, a joint winner of the 1963 Nobel Prize for physics, published a paper with the title *On the Unreasonable Effectiveness of Mathematics in the Natural Sciences* in Communications in Pure and Applied Mathematics. Wigner follows up on Galileo's idea that *The book of nature is written in the language of mathematics*. The title captures Wigner's main message in the following two key points:

(1) Mathematics is "effective" in describing phenomena in mechanics and physics.

(2) The "effectiveness" appears to be "unreasonable" or mysterious.

The title also suggests that the "effectiveness" of mathematics outside the natural sciences may be far less obvious.

We may view (1) as the main motivation to include mathematics as a basis in education in the natural sciences. We now present some key examples which would seem to give an affirmation of (1), and we then return to discuss (1) and (2). In particular, we seek to uncover the paradox of (2) following Wittgenstein's idea that an apparent paradox must be the result of a confusing use of language.

9.1   Newton's Model of Gravitation

The basic example is of course *Newton's theory of gravitation* based on (i) *Newton's (Second) Law of Motion*: $ma = F$, stating that the acceleration a of a body multiplied by its mass m is equal to the force F acting on the body, combined with (ii) *Newton's Inverse Square Law*, stating that the gravitational force F between two point masses is proportional to their masses m_1 and m_2 and inversely proportional to their distance r squared, with G being a universal constant of gravitation, that is $F = G\frac{m_1 m_2}{r^2}$. Combining these two laws Newton' s theory takes the form of a set of differential equations, which describes the motion of any system of point masses interacting by gravitational forces, which we may refer to as *Newton's model.* The theory extends to systems of (homogeneous) spheres with r being the distance between the centers of the spheres, and thus applies in particular to our *solar system* consisting of one (big) sun and 9 (small) planets, as well as a large number of moons.

Newton gave no explanation for the nature of gravitational forces, nor the form the Inverse Square Law, for which he was criticized by Leibniz. Maybe Leibniz was too tough: Still today, the nature of gravitational forces and their "action at a distance" is unknown. It is conjectured that the gravitational forces result from the exchange of certain "particles" called "gravitons", but little is known about the nature of such particles, or even that they "exist". *In my view they do not exist.*

Newton's model gives a very concise description of any system of bodies interacting through gravitation, but there is one catch: the differential equations are very difficult to solve by analytical mathematics expressing the solution in an analytical mathematical formula. It is only for the very special case of the *2-body problem* that we can find such a formula, which turns out to represent an *ellipse*. But already the 3-body problem presents unsurmountable analytical mathematical difficulties. *And it always will.*

Newton derived the solution of the 2-body problem in his *Principia Mathematica*, and thereby confirmed the laws discovered experimentally by Kepler. The success with the 2-body problem rocketed Newton to instant fame, and gave mathematics an enormous boost. It appeared that Man using mathematics could take up a competition with God and now, with no more limits to human understanding of the world, the industrial revolution could get started. The paradigm of our time is largely the same.

The fact that neither Newton, nor anybody else, could tackle even the 3-body problem analytically, did not take away the enthusiasm. The reason was that by a clever use of the 2-body solution one could find approximate solutions to e.g. our own planetary system, by first neglecting the interaction between planets and then correcting for this effect. In his monumental *Mecanique Celeste* Laplace made extensive computations of this form.

Today, with the computer, the differential equations of Newton's model are routinely solved computationally. Not by using the analytical 2-body so-

[handwritten margin note, left side:] point masses are a (useful) fiction

lution, but by directly solving the differential equations in each particular case. In principle this is done in a step-by-step procedure marching forward in time using small steps: In each time step the gravitational forces are computed from the present given positions of the bodies. Then the accelerations for all bodies are computed, then the velocities and finally the new positions are computed and used as input for the next time step. This procedure is referred to as solving the differential equations by a "time-stepping" method. In a certain sense computational solution of the n-body problem is thus "easy", at least in principle. But in practice it may require a lot of computations, and the computational work quickly grows with increasing n: in each time step we have to compute the forces between all the bodies (there are about $n^2/2$ such forces) and update acceleration, velocity and position for each body, and then we have to take many time steps. Thus, if n is large, e.g. $n = 10^6$, then a large computer is needed.

For planetary systems, the number of bodies may not be so large, but in *molecular dynamics* (see below) large numbers of molecules/atoms interact in a Newton-type model, which thus represents a real computational challenge. Further, in simulations of the *formation of galaxies*, large numbers of stars interact by gravitation. So there are many n-body problems demanding massive computations, but the rapid increase of computational power available to many quickly expand the scope of n-body simulations.

9.2 Laplace's Model of Gravitation

In *Mecanique Celeste* Laplace formulated a differential equation satisfied by the *gravitational potential* corresponding to a certain distribution of mass in space, which is referred to as *Laplace/Poisson's equation* involving *Laplace's differential operator* acting on the potential with the mass density as a given right hand side. By solving Laplace/Poisson's equation for the potential, one gets the gravitation force field as the *gradient* of the potential. Laplace could this way show that the gravitational force generated by a point mass satisfies the Inverse Square Law. So Laplace could give a "proof" of the Inverse Square Law, which Newton could not, but to do this Laplace had to assume (without being able to give a "proof") that a gravitational potential must satisfy Laplace/Poisson's equation. How daring!

9.3 Molecular Dynamics

Molecular dynamics gives a model for the interaction of molecules which is similar to Newton's Model for gravitation, with different intermolecular forces. Here the number of molecules may be large (remember that 1 mole consists of 10^{23} molecules) so even a small volume will contain a large number of molecules. Molecular dynamics offers new tools of simulating *protein folding* which is a basic process of *life*, but presents formidable challenges to modern computers. Even the molecular dynamics of ordinary *water*, which serves as the environment for *life processes*, remains a challenge.

9.4   Einstein's Law of Gravitation

Einstein presented an alternative to Newton's Theory of gravitation in *Einstein's differential equation* stating that the "curvature of space-time" is proportional to the mass density, and that the motion of a body under gravitation follows a "shortest path geodesic" in curved space-time in a "free fall".

Nota bene!

Einstein gave no explanation why the presence of mass would make space-time curved, nor why a body would necessarily follow a geodesic.

A few analytical solutions to Einstein's equation are known, but even today computational solution remains an outstanding challenge. For example, nobody has been able to compute the interaction of two black holes, which could give insight to the nature of the supposed associated *gravitational waves* and thus help experimental detection.

9.5   The Navier–Stokes Equations for Fluid Dynamics

The Navier–Stokes differential equations formulated 1821–45 describe the motion of a rich variety of incompressible fluid based on Newton's Second Law together with a constitutive equation stating that the shear forces are proportional to the strain velocity (assuming the fluid to be Newtonian). The Navier–Stokes equations are similar to a Newton model viewing a fluid to consist of many "fluid particles" which interact by pressure and viscous forces. The Navier–Stokes equations express the basic laws of mechanics of conservation of mass and momentum.

There are only a few known analytical solutions to the Navier–Stokes equations, and then only for very simple cases. In general solutions show features of *turbulent* rapidly changing flow in time and space. Computational solution of the Navier–Stokes including turbulent flow remains a main challenge today. We will come back to this topic below.

9.6   Schrödinger's Equation

Modern *Atom Physics* is based on *Schrödinger's equation* first presented in 1925, which describe *Quantum Physics*. Schrödinger's equation may be viewed as a generalization of classical Newtonian mechanics, where *momentum* is replaced by the Laplace operator acting on a *wave function* interpreted as a probability that a particle will be in a certain region of space at a certain time. Schrödinger formulated his equation so as to admit certain wave-like solutions that he liked but did not even attempt to try to "prove" the validity of his equations, but they seem to give precise predictions in many cases.

Again these differential equations allow analytical solution only in a few very simple cases such as the Hydrogen atom.

Schrödinger's equation is particularly demanding computationally since the number of space dimensions is equal to $3N$ where N is the number of electrons/kernels. Thus already one atom with many electrons poses a very demanding computational problem. Modern Atom Physics and *Physical Chemistry* can largely be described as the science of solving Schrödinger's equation approximately by different approaches. The Nobel Prize of 1998 was awarded for computational methods for solving Schrödinger's equation, based on replacing a large number of electrons by a single *electron density* thus reducing the number of space dimensions to the usual 3.

9.7   Discussion of Effectiveness

The mathematical models presented above may be viewed as key examples of the "effectiveness" of mathematics in mechanics, physics and chemistry. Newton's, Navier–Stokes, Einstein's and Schrödinger's equations are all differential equations which "describe" fundamental aspects of the world. More precisely, the equations express in concise mathematical form certain basic principles of physics. For example, Newtons equations combine Newton's Second Law $ma = F$ with the Inverse Square Law $F = Gm_1m_2/r^2$, and Navier–Stokes equations express Newton's Second Law together with an equation coupling viscous forces to velocity gradients.

Undoubtedly, all these mathematical models may be viewed as being extremely "effective" in the sense that the models can be specified on a couple of lines and yet seem to describe a very wide variety of different physical scenarios. One may express this as a (remarkable) "effectiveness in formulation". The extreme variant of this point of view is the current quest for a "basic equation for everything" combining the 4 basic forces of physics (strong and weak interactions, electromagnetic and gravitational forces) in one single model of string theory.

But there is a hook, a serious hook: all the equations are, except in very few special cases, impossible to solve by analytical mathematical methods. Apparently, we are thus able to write down equations which seem to describe physical realities, but we seem to be unable to solve the equations, at least by analytical mathematical methods. Of course this may make us question the "effectiveness", since after all it is solutions that we are after. We may ask what the value of a mathematical model in the form of a differential equation may be, if we cannot solve the differential equation, except in a few very special cases. The main use of a mathematical model is ultimately to make predictions, and predictions are made by solving the equations of the model.

The net result is that we may seriously question the "effectiveness" of the presented mathematical models, which usually are presented as key examples of the "effectiveness of mathematics in the natural sciences". Apparently, the models are "effective" from the point of view of economy of formulation as a certain set of differential equations expressing basic laws of physics, but the models are not very "effective" from the point of view of actually solving the equations to get solutions.

So, maybe the "effectiveness of mathematics" is largely an illusion. If so, the paradox and mystery of the "unreasonable effectiveness" would simply disappear.

We sum up: If we have a phenomenon which can be described by some basic laws, then there is a chance that we can find a corresponding mathematical model expressing the basic laws typically as a set of differential equations. There are many such phenomena, but there are also many phenomena where the basic laws seem to be missing or not fully known. This is often the case in social sciences, and this is the reason why mathematics is not considered to be very "effective" in these areas.

To give some further light on this aspect we take a look at a mathematical model used routinely all over the world with results presented each day along with the news on television.

9.8 Weather Prediction

We are all familiar with the weather report on TV: A prediction of the development of the weather in some area for some period of time is made, and usually the prediction is displayed in a movie showing the evolution of zones of high and low pressure along with the wind speed, the expected amount of rain or snow together with the variation of temperatures. The time scale may run from hours for local reports around an airport, to a day for a region, to long time predictions of e.g. global warming. But how is the simulation made? Well, of course by solving a set of differential equations supposedly modeling the atmosphere. Some of the differential equations in the model express basic conservation laws (of e.g. mass and momentum), while others are constitutive equations modeling for example the effect of clouds on heat radiation. We may know the basic conservation laws, but the constitutive equations have coefficients which we have to determine.

[handwritten margin note, left side: What is this constitutive?]

[handwritten note, bottom right: by experiment? by measuring?]

Thus, in many cases we may not know all the equations we want to solve in a simulation, so a large part of the effort will have to go into first determining what equations to solve, typically determining various coefficients, like heat conductivity or viscosity. Again this may be approached by computation, where we compute with a certain set of coefficients, compare the corresponding computed solution to observations, and then change the coefficients to get a better match. *Recalibrate the model*

So, also in science, we may have to struggle hard (compute) to find the equations to solve in a simulation. The "effectiveness in formulation" may then become as questionable in natural science as in economy or social science. *economic*

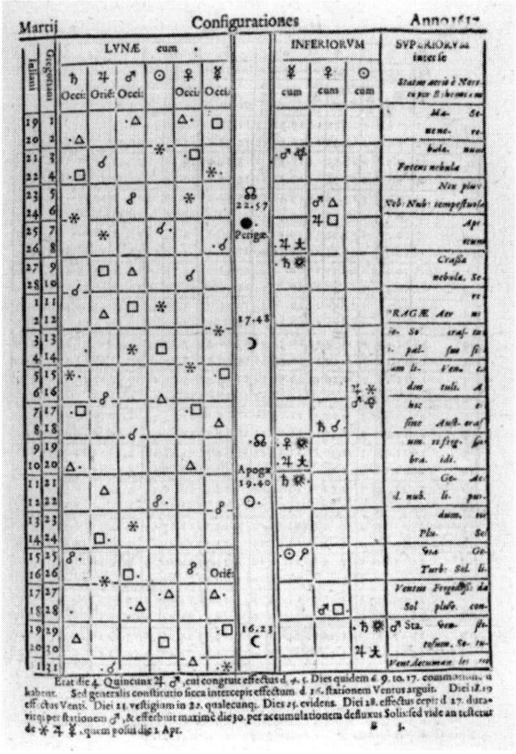

FIGURE 9.1. Table from Kepler's *Ephemerides*: Kepler speculated that the weather was affected by planetary influences. *I'm sure it is.*

10
Do We live in "The Best of Worlds"?

The Best World has the greatest variety of phenomena regulated by the simplest Laws. (Leibniz)

Newton's classical mechanics, Schrödinger's quantum mechanics, Navier–Stokes fluid mechanics and Einstein's cosmology are all examples of mathematical models in the form of differential equations expressing basic laws such as conservation of mass, momentum or energy.

Each differential equation describes a physical world on a certain *scale* ranging from the atomic scale of Schrödinger's equation to the cosmological scale of Einstein's equation.

Each equation may be viewed as a wonderful example of the "unreasonable effectiveness of mathematics in the natural sciences" capturing a large variety of phenomena in one differential equation expressing a basic principle. Each equation may be taken as evidence that we live in the "Best of Worlds" following the idea of Leibniz that a best possible world would be one of maximal complexity governed by the simplest laws.

The genetic code seems to support Leibniz idea: the complete design of a living being, from the amoeba to homo sapiens is encoded in a genome consisting of a set of chains of amino acids contained in the kernel of each cell. This startling idea of the Creator is now being uncovered: the recording of the human genome was completed in 2000 and its functionality is now (slowly) being uncovered. The process is like first copying a very thick book in a completely unknown language about a completely unknown subject (which may be quite easy with a good copying machine), and then trying

FIGURE 10.1. The first page of the manuscript published in 1684, where Leibniz introduced calculus to describe basic aspects of his "Best of Worlds".

to understand what the book says (which could be infinitely much more difficult).

Also Stephen Wolfram, the creator of the mathematical software *Mathematica*, follows the line of thought of Leibniz in his latest book *A New Kind of Science*. Wolfram there plays with small computer codes expressing some simple law of interaction which generate interesting patterns, like primitive forms of life.

Going back to mechanics and physics, the worlds of Newton, Schrödinger, and Einstein thus appear to be "Best Possible", in a way. But we know that there is a catch, a hook: If the corresponding world has maximal complexity, then solutions to the wonderful equations of Newton, Schrödinger and Einstein may be very complex. And that is the catch: a complex solution is not easy to capture in an analytical mathematical formula. So with only analytical mathematics as a tool we get quickly stuck: We simply can't solve the equations, and their secrets remain closed to us.

Yes, of → So what is then the use of an equation, to which we can't find solutions?
what use? It is like a riddle without an answer. Maybe amusing, but what is the use of it?

Or, is it indeed possible to find some kind of solutions? Like in our lives: It does not seem easy to tell beforehand what expression our genome will take as if we had an analytical solution given as a birth present (which our insurance company certainly would like to have), but nevertheless somehow we seem to live our lives one way or the other, day after day. Or the other way around: maybe it is the genome that lives its life and our own life is some part of it, which we cannot understand.

11
The Reasonable Effectiveness of Computational Mathematics

When, several years ago, I saw for the first time an instrument which, when carried, automatically records the number of steps taken by a pedestrian, it occurred to me at once that the entire arithmetic could be subjected to a similar kind of machinery so that not only addition and subtraction, but also multiplication and division could be accomplished by a suitably arranged machine easily, promptly and with sure results... For it is unworthy of excellent men to lose hours like slaves in the labor of calculations, which could safely be left to anyone else if the machine was used... And now that we may give final praise to the machine, we may say that it will be desirable to all who are engaged in computations which, as is well known, are the mangers of financial affairs, the administrators of others estates, merchants, surveyors, navigators, astronomers, and those connected with any of the crafts that use mathematics. (Leibniz)

So there we stand and look at the equations, and don't know how to solve them. Or do we know? Yes, of course we can always try to compute the solutions. Let's see what we can do.

We thus ask the question if we can compute solutions to the equations of Newton, Schrödinger and Einstein? Can we do that? Yes, and no!

For some problems, we already have efficient methods to compute solutions, for other problems we may expect to develop new methods that would produce solutions, while other problems seem very difficult or impossible to solve computationally. We may phrase this as "the reasonable effectiveness of computational mathematics".

Thus: computational mathematics is reasonable in the sense that we can solve reasonably many problems with a reasonable amount of computational work. That is, we cannot solve all problems using computational mathematics (that would have been unreasonable), but some problems (which is more reasonable).

Let's give some key examples (to which we will return in more detail below for the reader who wants to know a bit more).

11.1   The Solar System: Newton's Equations

The evolution of our solar system is accurately described by Newton's equation. Knowing the positions of all the planets and their moons at a given moment (say time $t = 0$ which may be today), we can compute the positions for some length of time ahead (say for $t \leq T$ where T is a final time) by solving Newton's equations by time stepping: we start knowing the initial positions at $t = 0$ and compute their positions at time $t = k$, where k is a positive *time step* like 1 hour, 1 day or even 1 year. We then get new initial positions at $t = k$ and repeat the procedure to get new positions at time $t = 2k$, and so on. We would then advance our solar system like a big Clock, with each time step being a "tic" by the Clock.

But, in each time step we would make a small error in the computation; we cannot compute the positions exactly, since we are using finite precision (finite number of decimals) in the computation, and since the computational method also adds an error (which would be non-zero even if we could compute with infinite precision). So in each time step we make a little error and after many time steps, all the little errors may add up to a big error, and the computation gives completely false results. We also have errors from data; from errors in measured initial positions, from errors in estimated values of the masses of the planets and moons, and also an error from a lack of precise knowledge of the universal gravitational constant G entering the equations. Altogether, we have errors from discretization (time stepping), round-off (finite precision arithmetics), and data.

When we solve Newton's equations, we notice that different celestial objects show different error growth: For example, to computationally predict the position of our moon over a period of more than thousand years would require higher accuracy of data than the current knowledge of G up to 5-6 decimal places. On the other hand, we would see smaller error growth in the position of e.g. Pluto. So, we may predict the evolution of the solar system over a period of thousands or million years, depending on what information or *output* from the computations we want.

So, solving Newton's equations computationally is "Reasonably Efficient" but not more.

11.2   Turbulence: Navier–Stokes Equations

Why is it a mystery?

Turbulence is viewed as one of the main mysteries of Newtonian Mechanics. We know (firmly believe) that the motion of an incompressible viscous flow is governed by Navier–Stokes equations which express Newton's Second Law (conservation of momentum) and incompressibility (conservation of mass). If the viscosity is small (the Reynolds number is large), then we expect to see turbulent (highly fluctuating) solutions to Navier–Stokes equations. If we could solve the equations analytically, we could uncover the secret of turbulence, but we can't: only very few very simple (non-turbulent) analytical solutions are know (and these are probably all that ever may be known).

HA!

Again we have to resort to computational methods: And yes, we can solve the Navier–Stokes equations to a certain extent computationally. There are aspects of turbulent flow (outputs), which are computable and other aspects which appear to be "uncomputable". For example, we may compute the *drag coefficient* c_D of our car, which is a mean value in time of the momentary *drag force* $D(t)$, which is the total force from the air acting on the body of the car at time t. But we may not compute $D(t)$ accurately at a given time t, because $D(t)$ is rapidly fluctuating. Neither can we measure $D(t)$ and thus we get the message that this information is forever hidden to us. But we can accurately compute the mean value in time of $D(t)$ which is the drag coefficient c_D. We give more details below.

[coefficient of drag

So again, computational solution of the Navier–Stokes equations is reasonable in the sense that we can compute some outputs of turbulent flow (typically certain mean values) but not point values in space or time.

22 Oct '05

Mean means average. I would argue that there are no point values in space or time and that this is a fallacy of The Calculus. Further I would

11.3   Schrödinger's Equation

The Schrödinger equation is the basic equation of Quantum Mechanics and differs essentially from e.g. Navier–Stokes equations. First of all the unknown is a certain *wave function* interpreted as the *probability* of a certain distribution of atoms and electrons. Second, the number of space dimensions is equal to $3N$ where N is the number of electrons. The idea is that in a sense each electron has its own copy of three dimensional space and we then get an incredibly rich world of high dimension. The result is that only the very simplest case of the Hydrogen atom with *one* electron allows an analytical solution. *So there!*

Eek!

Since we have to discretize in space to solve the Schrödinger equation computationally, we also quickly get drowned by the large number of dimensions: we quickly fill up even the largest computer and get nowhere. The only chance is to come up with clever computational methods where we do not seek to follow each individual electron but rather work with one

argue that space and time do not consist of points but of → [a wave that extends...]

single *electron density*. The difficulty with this approach is that we don't have an equation to solve for the electron density, but first have to construct the equation to solve. *Equation first, electron second.*

So either, we have an equation (Schrödinger's equation), which we cannot solve computationally, or we aim for a solvable equation, which we however don't know what it could be. As we said, to find the equations to solve computationally itself may come out as a result of a computational procedure: To compute the solution of an equation, we first have to compute the equation! Again, we may view this as something in fact "reasonable". It would be unreasonable to expect that the equation was given to us by God, and our task was "just" to solve it. Life is not simple: It is not just to "get married and be happy", we first have find someone to marry.

A truly vicious circle.

11.4   Einstein's Equation

Einstein's equation describing the large scales of our Universe with *gravitation* as the main force, turns out to be very much more difficult to solve computationally than the corresponding simpler model in the form of Laplace's equations. And of course as usual, analytical solution is impossible except in a few simple situations. As of now, nobody seems to be able to computationally solve Einstein's equation accurately in any generality. But presumably it will be possible to get around the difficulties. A good aspect is that the dimension is 4 (space and time), so it should be possible to discretize. The Main difficulties seem to come from a lack of understanding what the equations really mean and how they should be formulated in a computational approach. Also, what initial and boundary conditions to pose present difficulties.

Following our idea that computational mathematics is reasonably efficient, we should expect some progress soon concerning computational solution of Einstein's equation. This could be expected to give new insights concerning black holes and gravitational waves, which physicists believe exist but which are difficult to observe.

So, in a sense, Einstein's equation is a fraud.

11.5   Comparing Analytical and Computational Solution Techniques

In Chapter 12 we touch the difference between classical music, where the set of notes to be played is given, and jazz music where improvisation is an important feature. Many analytical solutions are like the wonderful inventions by Bach, something that we (with a lot of practice) can reproduce more or less successfully (or even "interpret" if we are professionals), but which we cannot write down ourselves.

closed?

Analytical solution is thus "closed" in the sense that normally we can only hope to copy what someone more clever than ourselves has already achieved. Calculus books are filled with "tricks", which the teacher has learnt to master by teaching the same course many times. It is like following an already existing prepared path in the jungle. This makes the teacher very important as the source to the secret of where to go next, following the prepared path (which we don't see very well). The student thus gets a passive role, and his own initiatives usually lead astray: the problem is to find the next sign dropped by the teacher indicating where to go.

The basic reason is that analytical solutions are sparse and difficult to find. Unless you carefully set up the problem just right, no one will be able to find a solution. So analytical solutions may be viewed as valuable gems, which are not easy to find.

On the other hand, by computation you can solve almost any problem, more or less, and with more or less effort. And you don't have to get stuck completely: you have your good friend the computer which can help you. To produce an analytical solution you just have your brains and paper and pencil, so you are alone.

Thus, computational mathematics is "open" in the sense that whatever problem you approach you can at least do something. Like in jazz music: if you only know (or feel) the chords, you can make some music, even if it is not on the level of Bach (or a Master of Improvisation like Keith Jarrett.)

Of course, viewing mathematics as a training ground for problem solving, it may be very rewarding to learn how to use computational techniques, because you don't get stuck as much, and always getting stuck does not build confidence (very important in problem solving), but the contrary (which is very destructive).

11.6 Algorithmic Information Theory

Suppose you compare the length of a computer program with the "information" that may come out by running it on the computer. The computer program could be one of Stephen Wolfram's little codes that produce complex patterns, or an implementation of a computational method for solving the Navier–Stokes equations, which produces a turbulent complex solution. In both cases the computer program could be very short. In the Navier–Stokes case this would reflect that the Navier–Stokes equations themselves can be written down very concisely (2 lines) and that it is possible to solve them computationally using a numerical method that also can be written down in a few lines. Maybe not the most efficient program, but still capable of producing a complex solution.

So in both cases, the computer program may be very short, but the result that comes out very long in the sense that it takes a lot of memory to store the entire evolution in space and time of the solution. The program may thus be a few lines long, while the solution may take 2 Gb to store. And of course, to produce the solution we have to put in a lot of computational work.

This is similar to the genetic code (see below), which is quite short in terms of storage, while the entire life that may come out from the genetic code may be very rich and would require a lot of storage to record. And of course to express the genetic code in the form of real life, a lot of work would have to be added (and energy consumed).

If the code is short but the result is long, we may say that there is some structure or order (Life is short, Art is long). If the code is as long as the result, this may be viewed as evidence of *randomness*.

So the Best World may be viewed as a complex world regulated by a short computer program. Viewing the Creator of such a Best World as a programmer, this would indicate that the Creator is a good programmer, with access to a lot of computational resources.

This also couples to the concept of *depth* of information, where the depth measures the amount of work required to produce a certain *result* or *answer*. The answer may be short like a "no" (or "42") as a result of a very long thought process (or computation), e.g. in a proposal to marriage, and may thus be of considerable depth. Or a very short one of little depth as a response to the question "Do you still smoke?" or "Is two plus two equal to five?". One may say that Fermat's Last Theorem has a considerable depth, taking the length of its proof as a measure, with a very short answer.

Further, if a short computer program is capable of producing a complex output when executed on some short input data, it may be used to transmit the output data in an efficient way. Instead of first computing the output at location A and then sending the output to location B, which may require sending very large amounts of data, one would send the short program together with the short input from A to B, and then execute the program at B to produce the output.

11.7   How Smart is an Electron?

Somehow each electron in a molecule has to find out what to do, by somehow "solving" its own copy of the Schrödinger equation. So "solving" in some sense has to be "simple", since after all a single electron must have a limited intelligence. This indicates that there may be a simple computer program for the computational solution of Schrödinger's equation.

Arguing this way, the world as it evolves, seems to find its way in a massive "physical computation" with many interacting particles each one solving its own equation, which the physicists may say ultimately boils down to an "exchange of something". A process which maybe could be modeled in a digital computation with instead exchange of digits. The massiveness of the computation with many particles interacting would then be what opens to complexity. A short computer program applied in a massively parallel computation would produce the complex Best World.

11.8   The Human Genome Project

In 2003 the Human Genome Project was completed by obtaining the sequence of the 3 billion base pairs making up the human genome, or genetic code, distributed over about 30,000 genes. The order of the bases A, T, C and G spells out the exact instruction needed to maintain and reproduce a living organism, whether it is a human being, a tree or a microbe. Each cell of the organism has a copy of the complete genome. The genome can be viewed as a computer code which generates a certain living organism when the code is executed many times in many cells. The code is very compact (less than 1 Gb of storage), but its expression very complex, thus meeting Leibniz criterion of a Best World.

> Algorithmic information theory is the result of putting Shannon's information theory and Turing's computability theory into a cocktail shaker and shaking vigorously. The basic idea is to measure the complexity of an object by the size in bits of the smallest program for computing it. (G. J. Chaitin)

And do the virtual particles also exchange 2nd order virtual particles and so on ad infinitum?

12

Jazz/Pop/Folk vs. Classical Music

Music is the pleasure the human soul gets from counting, without knowing it. (Leibniz)

We present another analogy taken from the area of music. We know that there are many profound connections between music and mathematics going way back to the Pythagoreans in old Greece, who discovered the mathematics of musical scales and harmony as simple fractions such as $3/2$ and $9/4$.

We know that there is jazz/pop/folk music and classical music, which are quite different even if they share elements of musical scales/harmony and rhythm. In jazz music the basic structure of a tune being played is determined beforehand in the form of a basic chord progression and rhythmic pattern (e.g. a 12 bar blues pattern), but the details of the music (melody, harmony and rhythm) is created during performance in improvisations on the given harmonic/rhythmic structure. One may view the improvisation as a computation being performed during performance on a certain set of data with certain (partly random) decisions being taken as the improvisation develops.

In classical music on the other hand, the full score (set of notes) is set beforehand. The individual members of a symphony orchestra do not have the freedom to improvise and play what they feel to play for the moment (even if it would be harmonically and rhythmically correct), but will have to play the given notes. The conductor has a freedom of interpretation in tempo and phrasing which may be different from one performance to the

—partiture

next according to inspiration. But the given set of notes have to be played according to the partiture created by the composer.

We may view computational mathematics as sharing a quality of jazz music: a computation follows an algorithm which prescribes a certain structure of the computation, like a chord progression, but each run of the computation may be done with new data and a new result will come out. The computation may be designed to simulate the evolution of the atmosphere to give input to a weather prediction report, and there may even be random elements in the computation reflecting lack of data or strong sensitivity to small disturbances.

On the other hand, the most essential element of pure mathematics as a science is considered to be a proof of a theorem created by an individual mathematician, which may be viewed as a kind of score created by an individual composer. Often the theorem is given a name *memorizing* the mathematician who first created a proof of that theorem. For example we have: Banach's Fixed Point Theorem, Brouwer's Fixed Point Theorem, Lefschetz' Fixed Point Theorem, Schauder's Fixed Point Theorem, Kakutani's Fixed Point Theorem, ...   *A lot of fixed point theorems,*

It appears that jazz/pop/folk music reaches many more people than classical music, which also get large audiences when the classics are performed but often very few when contemporary music is played. Analogously, we all meet many products of computational mathematics in our everyday life, while the activities of pure mathematics more and more seems to develop into activities understood by very small groups of specialists.

Of course, the analogy presented only captures, at best, some truth and does not describe the whole picture. Some of the reviewers of the book did not not understand the message of this chapter. How about you?

I understood it and its relevance.

FIGURE 12.1. Charlie Parker, a genius of improvisation.

13

The Right to Not Know

> To my knowledge the President had no prior knowledge of the bugging plan, and such knowledge by Nixon, would be difficult to believe. (Nixon's campaign Deputy Director Jeb Stuart Magruder, testifying under oath 1972)

To not know, to not be informed, is a trait of political life which we all have met. The Swedish Prime Minister did not know anything about bribes in *← Oh?* weapons deals between Sweden and India, which caused a severe government crisis in India. Nixon did not know anything about Watergate, Kohl did not know about the finances of his electoral campaigns, and Blair was not aware of missing proofs of the existence of weapons of mass destruction in Iraq. Or did they know? We don't know for sure.

Also in science, denial and ignorance may be a powerful tool. By simply ignoring new facts, an old paradigm may be upheld for yet some period of time. Ignorance is based on a (possibly silent) agreement to ignore. Like in the H.C. Andersen story about the Emperors New Clothes, where everybody agreed to ignore the fact that the Emperor was naked, except a little boy...

It appears that mathematicians have the right to be ignorant about mathematics didactics, and experts of mathematics didactics have the right to not know too much about contemporary research in mathematics. And both groups seem to have the right to not be well informed about computational mathematics. Similarly, the Mathematics Delegation seems to have the right to pretend that computational mathematics does not exist.

But there are hooks of course: to pretend to not know, while knowing, or to be truly ignorant, may be a risky tactic. In business, it would certainly be wiser to try to follow what the other actors on the market are doing, than to simply forget about their existence. And the same rule should apply to science: it would seem wiser to be informed about what your competitors are doing, than to know nothing. At best this knowledge could give you the good feeling of knowing that you are ahead, and if you see that you are behind, you'd better catch up.

In an open society with freedom of speech, there are limits to what can be ignored: there are digging journalists and dogged scientists, which will simply not disappear even if ignored for a long time.

We have written several debate articles in the Swedish daily press about the need of reform in mathematics education. We have met very positive response from many, but none at all from pure mathematics and mathematics didactics. In our last debate article in *Göteborgs-Posten* on March 6 2004 (see Appendix) we suggested that the lack of reaction could reflect a lack of knowledge of computational mathematics, i.e., a genuine lack of ideas of how to respond to what we were saying. And we did not get any reaction indicating that this was wrong.

⌐Exactly!?

14
An Agenda

... there is no study in the world which brings into more harmonious action all the faculties of the mind than [mathematics], ... or, like this, seems to raise them, by successive steps of initiation, to higher and higher states of conscious intellectual being... (Sylvester 1814–1897)

We propose the following agenda:

- Create a scientific basis of mathematics including both pure and computational mathematics.

- Improve the interaction between pure/computational mathematics and didactics of mathematics.

- Develop reformed mathematics educations on all levels based on the new scientific basis.

Needless to say, this agenda is demanding and would need contributions from many to get an impact. Our Body&Soul Project represents a small initial effort in this direction.

The agenda is short, and it should be expanded and be made more detailed. Here we only give a few comments:

14.1 Foundations of Computational Mathematics

The famous mathematician Stephen Smale, Fields Medalist 1966, has initiated the formation of the organization *Foundations of Computational*

Mathematics FoCM which "supports and promotes research on foundations of computational mathematics, and fosters interaction among mathematics, computer science and other areas of computational science through its conferences, workshops and publications". The FoCM series of conferences shows a development from an initial enthusiasm (FoCM 95) to identify the "foundations", to a realization (FoCM 02) that it was difficult to really properly identify them.

Our own experience from our research and from writing several books, and from participating in FoCM 95/02, is that the foundations of computational mathematics are the same as those of constructive calculus as presented e.g. in Body&Soul. Thus, if we take away the non-constructive aspects of calculus, then what remains is constructive calculus which together with the computer gives us computational mathematics. This means that among the foundations of computational mathematics we find computer arithmetic with real numbers, Banach's fixed point theorem, Newton's method, Euler's method, Gaussian elimination, the Conjugate Gradient method and Galerkin's method.

14.2   New Possibilities for Mathematics Education

We have remarked above on, from our perspective, a missing interaction between mathematics didactics and computational mathematics. We believe that computational mathematics, or constructive calculus, opens new possibilities in the teaching of mathematics, and thus should open a new interesting and fruitful field of mathematics didactics. Until now we have seen little exploitation of these new possibilities. We hope that this little book can help to stimulate an interest from mathematics didactics in computational mathematics.

Part II

Essence

15

A Very Short Calculus Course

Mathematics has the completely false reputation of yielding infallible
conclusions. Its infallibility is nothing but identity. Two times two is
not four, but it is just two times two, and that is what we call four
for short. But four is nothing new at all. And thus it goes on in its
conclusions, except that in the height the identity fades out of sight.
(Goethe)

15.1 Introduction

Following up on the general idea of science as a combination of formulating
and solving equations, we describe the bare elements of this picture from
a mathematical point of view. We want to give a brief glimpse of the main
themes of Calculus that will be discovered if we work through the volumes of
Body&Soul. In particular, we will encounter the magical words of *function*,
derivative, and *integral*. If you have some idea of these concepts already,
you will understand some of the outline. If you have no prior acquaintance
with these concepts, you can use this section to just get a first taste of
what Calculus is all about without expecting to understand the details at
this point. Keep in mind that this is just a glimpse of the actors behind
the curtain before the play begins!

We hope the reader can use this chapter to get a grip on the essence of
Calculus by reading just a couple of pages. But this is really impossible in
some sense because calculus contains so many formulas and details that it
is easy to get overwhelmed and discouraged. Thus, we urge the reader to

browse through the following couple of pages to get a quick idea and then return later and confirm with an "of course".

On the other hand, the reader may be surprised that something that is seemingly explained so easily in a couple of pages, actually takes several hundred pages to unwind in this book (and other books). We don't seem to be able give a good explanation of this "contradiction" indicating that "what looks difficult may be easy" and vice versa.

15.2   Algebraic Equations

There are *algebraic equations* of the form: find $\bar{x}$ such that

$$f(\bar{x}) = 0, \tag{15.1}$$

where $f(x)$ is a *function* of x. Recall that $f(x)$ is said to be a function of x if for each number x there is a number $f(x)$ assigned.

We call $\bar{x}$ a *root* of the equation $f(x) = 0$ if $f(\bar{x}) = 0$. The root of the equation $15x - 10 = 0$ is $\bar{x} = \frac{2}{3}$. The positive root $\bar{x}$ of the equation $x^2 - 2 = 0$ is equal to $\sqrt{2} \approx 1.41$. There are different methods to compute a root $\bar{x}$ satisfying $f(\bar{x}) = 0$ such as the Bisection Method and Newton's Method.

15.3   Differential Equations

We will also consider the following *differential equation*: find a function $x(t)$ such that for all t

$$x'(t) = f(t), \tag{15.2}$$

where $f(t)$ is a given function, and $x'(t)$ is the *derivative* of the function $x(t)$. This equation has several new ingredients. First, we seek here a *function* $x(t)$ with a set of different values $x(t)$ for different values of the variable t, and not just one single value of x like the root the algebraic equation $x^2 = 2$ considered above. $x = x(y) = \frac{y}{15}$. Secondly, the equation $x'(t) = f(t)$ involves the derivative $x'(t)$ of $x(t)$, so we have to investigate derivatives.

A basic part of Calculus is to (i) explain what a derivative is, and (ii) solve the differential equation $x'(t) = f(t)$, where $f(t)$ is a given function. The solution $x(t)$ of the differential equation $x'(t) = f(t)$, is referred to as an *integral* of $f(t)$, or alternatively as a *primitive function* of $f(t)$. Thus, a basic problem of Calculus is to find a primitive function $x(t)$ of a given function $f(t)$ corresponding to solving the differential equation $x'(t) = f(t)$.

We now attempt to explain (i) the meaning of (15.2) including the meaning of the derivative $x'(t)$ of the function $x(t)$, and (ii) give a hint at how to find the solution $x(t)$ of the differential equation $x'(t) = f(t)$ in terms of the given function $f(t)$.

As a concrete illustration, let us imagine a car moving on a highway. Let t represent *time*, let $x(t)$ be the *distance* traveled by the car at time t, and let $f(t)$ be the *momentary velocity* of the car at time t, see Fig. 15.1.

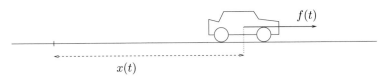

FIGURE 15.1. Highway with car (Volvo?) with velocity $f(t)$ and travelled distance $x(t)$.

We choose a starting time, say $t = 0$ and a final time, say $t = 1$, and we watch the car as it passes from its initial position with $x(0) = 0$ at time $t = 0$ through a sequence of increasing intermediate times $t_1, t_2,...$, with corresponding distances $x(t_1)$, $x(t_2)$,..., to the final time $t = 1$ with total distance $x(1)$. We thus assume that $0 = t_0 < t_1 < \cdots < t_{n-1} < t_n \cdots < t_N = 1$ is a sequence of intermediate times with corresponding distances $x(t_n)$ and velocities $f(t_n)$, see Fig. 15.2

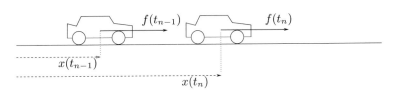

FIGURE 15.2. Distance and velocity at times t_{n-1} and t_n.

For two consecutive times t_{n-1} and t_n, we expect to have

$$x(t_n) \approx x(t_{n-1}) + f(t_{n-1})(t_n - t_{n-1}), \qquad (15.3)$$

which says that the distance $x(t_n)$ at time t_n is obtained by adding to the distance $x(t_{n-1})$ at time t_{n-1} the quantity $f(t_{n-1})(t_n - t_{n-1})$, which is the product of the velocity $f(t_{n-1})$ at time t_{n-1} and the *time increment* $t_n - t_{n-1}$. This is because

change in distance $=$ average velocity $\times$ change in time,

or traveled distance between time t_{n-1} and t_n equals the (average) velocity multiplied by the time change $t_n - t_{n-1}$. Note that we may equally well connect $x(t_n)$ to $x(t_{n-1})$ by the formula

$$x(t_n) \approx x(t_{n-1}) + f(t_n)(t_n - t_{n-1}), \qquad (15.4)$$

corresponding to replacing t_{n-1} by t_n in the $f(t)$-term. We use the approximate equality $\approx$ because we use the velocity $f(t_{n-1})$ or $f(t_n)$, which is not exactly the same as the average velocity over the time interval from t_{n-1} to t_n, but should be close to the average if the time interval is short (and the velocity does not change very quickly).

EXAMPLE 15.1. If $x(t) = t^2$, then $x(t_n) - x(t_{n-1}) = t_n^2 - t_{n-1}^2 = (t_n + t_{n-1})(t_n - t_{n-1})$, and (15.3) and (15.4) correspond to approximating the average velocity $(t_n + t_{n-1})$ with $2t_{n-1}$ or $2t_n$, respectively.

The formula (15.3) is at the heart of Calculus! It contains both the derivative of $x(t)$ and the integral of $f(t)$. First, shifting $x(t_{n-1})$ to the left and then dividing by the time increment $t_n - t_{n-1}$, we get

$$\frac{x(t_n) - x(t_{n-1})}{t_n - t_{n-1}} \approx f(t_{n-1}). \tag{15.5}$$

This is a counterpart to (15.2), which indicates how to define the derivative $x'(t_{n-1})$ in order to have the equation $x'(t_{n-1)}) = f(t_{n-1})$ fulfilled:

$$x'(t_{n-1}) \approx \frac{x(t_n) - x(t_{n-1})}{t_n - t_{n-1}}. \tag{15.6}$$

This formula says that the derivative $x'(t_{n-1})$ is approximately equal to the *average velocity*

$$\frac{x(t_n) - x(t_{n-1})}{t_n - t_{n-1}}.$$

over the time interval between t_{n-1} and t_n. Thus, we may expect that the equation $x'(t) = f(t)$ just says that *the derivative $x'(t)$ of the traveled distance $x(t)$ with respect to time t, is equal to the momentary velocity $f(t)$.* The formula (15.6) then says that the velocity $x'(t_{n-1})$ at time t_{n-1}, that is the *momentary velocity* at time t_{n-1}, is approximately equal to the *average velocity* over the time interval (t_{n-1}, t_n). We have now uncovered some of the mystery of the derivative hidden in (15.3).

Next, considering the formula corresponding to to (15.3) for the time instances t_{n-2} and t_{n-1}, obtained by simply replacing n by $n-1$ everywhere in (15.3), we have

$$x(t_{n-1}) \approx x(t_{n-2}) + f(t_{n-2})(t_{n-1} - t_{n-2}), \tag{15.7}$$

and thus together with (15.3),

$$x(t_n) \approx \overbrace{x(t_{n-2}) + f(t_{n-2})(t_{n-1} - t_{n-2})}^{\approx x(t_{n-1})} + f(t_{n-1})(t_n - t_{n-1}). \tag{15.8}$$

Repeating this process, and using that $x(t_0) = x(0) = 0$, we get the formula

$$\begin{aligned} x(t_n) \approx X_n &= f(t_0)(t_1 - t_0) + f(t_1)(t_2 - t_1) + \cdots \\ &+ f(t_{n-2})(t_{n-1} - t_{n-2}) + f(t_{n-1})(t_n - t_{n-1}). \end{aligned} \tag{15.9}$$

EXAMPLE 15.2. Consider a velocity $f(t) = \frac{t}{1+t}$ increasing with time t from zero for $t = 0$ towards one for large t. What is the travelled distance $x(t_n)$ at time t_n in this case? To get an (approximate) answer we compute the approximation X_n according to (15.9):

$$x(t_n) \approx X_n = \frac{t_1}{1+t_1}(t_2 - t_1) + \frac{t_2}{1+t_2}(t_3 - t_2) + \cdots$$
$$+ \frac{t_{n-2}}{1+t_{n-2}}(t_{n-1} - t_{n-2}) + \frac{t_{n-1}}{1+t_{n-1}}(t_n - t_{n-1}).$$

With a "uniform" time step $k = t_j - t_{j-1}$ for all j, this reduces to

$$x(t_n) \approx X_n = \frac{k}{1+k}k + \frac{2k}{1+2k}k + \cdots$$
$$+ \frac{(n-2)k}{1+(n-2)k}k + \frac{(n-1)k}{1+(n-1)k}k.$$

We compute the sum for $n = 1, 2, .., N$ choosing $k = 0.05$, and plot the resulting values of X_n approximating $x(t_n)$ in Fig. 15.3.

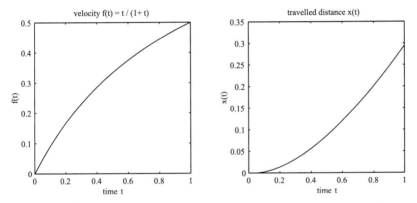

FIGURE 15.3. Travelled distance X_n approximating $x(t)$ for $f(t) = \frac{t}{1+t}$ with time steps $k = 0.05$.

We now return to (15.9), and setting $n = N$ we have in particular

$$x(1) = x(t_N) \approx f(t_0)(t_1 - t_0) + f(t_1)(t_2 - t_1) + \cdots$$
$$+ f(t_{N-2})(t_{N-1} - t_{N-1}) + f(t_{N-1})(t_N - t_{N-1}),$$

that is, $x(1)$ is (approximately) the sum of the terms $f(t_{n-1})(t_n - t_{n-1})$ with n ranging from $n = 1$ up to $n = N$. We may write this in more condensed form using the *summation sign* Σ as

$$x(1) \approx \sum_{n=1}^{N} f(t_{n-1})(t_n - t_{n-1}), \tag{15.10}$$

which expresses the total distance $x(1)$ as the sum of all the increments of distance $f(t_{n-1})(t_n - t_{n-1})$ for $n = 1, ..., N$. We can view this formula as a variant of the "telescoping" formula

$$
x(1) = x(t_N) \overbrace{-x(t_{N-1}) + x(t_{N-1})}^{=0} \overbrace{-x(t_{N-2}) + x(t_{N-2})}^{=0} \cdots
$$

$$
\overbrace{+ -x(t_1) + x(t_1)}^{=0} -x(t_0)
$$

$$
= \underbrace{x(t_N) - x(t_{N-1})}_{\approx f(t_{N-1})(t_N - t_{N-1})} + \underbrace{x(t_{N-1}) - x(t_{N-2})}_{\approx f(t_{N-2})(t_{N-1} - t_{N-2})} + x(t_{N-2}) \cdots - x(t_1)
$$

$$
+ \underbrace{x(t_1) - x(t_0)}_{\approx f(t_0)(t_1 - t_0)}
$$

expressing the total distance $x(1)$ as a sum of all the increments $x(t_n) - x(t_{n-1})$ of distance (assuming $x(0) = 0$), and recalling that

$$
x(t_n) - x(t_{n-1}) \approx f(t_{n-1})(t_n - t_{n-1}).
$$

In the telescoping formula, each value $x(t_n)$, except $x(t_N) = x(1)$ and $x(t_0) = 0$, occurs twice with different signs.

In the language of Calculus, the formula (15.10) will be written as

$$
x(1) = \int_0^1 f(t)\, dt, \tag{15.11}
$$

where $\approx$ has been replaced by $=$, the sum $\sum$ has been replaced by the *integral* $\int$, the increments $t_n - t_{n-1}$ by dt, and the sequence of "discrete" time instances t_n, running (or rather "jumping" in small steps) from time 0 to time 1 corresponds to the *integration variable* t running ("continuously") from 0 to 1. We call the right hand side of (15.11) the *integral* of $f(t)$ from 0 to 1. The value $x(1)$ of the function $x(t)$ for $t = 1$, is the *integral* of $f(t)$ from 0 to 1. We have now uncovered some of the mystery of the integral hidden in the formula (15.10) resulting from summing the basic formula (15.3).

The difficulties with Calculus, in short, are related to the fact that in (15.5) we divide with a small number, namely the time increment $t_n - t_{n-1}$, which is a tricky operation, and in (15.10) we sum a large number of approximations, and the question is then if the approximate sum, that is, the sum of approximations $f(t_{n-1})(t_n - t_{n-1})$ of $x(t_n) - x(t_{n-1})$, is a reasonable approximation of the "real" sum $x(1)$. Note that a sum of many small errors very well could result in an accumulated large error.

We have now gotten a first glimpse of Calculus. We repeat: the heart is the formula

$$
x(t_n) \approx x(t_{n-1}) + f(t_{n-1})(t_n - t_{n-1})
$$

or setting $f(t) = x'(t)$,

$$x(t_n) \approx x(t_{n-1}) + x'(t_{n-1})(t_n - t_{n-1}),$$

connecting increment in distance to velocity multiplied with increment in time. This formula contains both the definition of the integral reflecting (15.10) obtained after summation, and the definition of the derivative $x'(t)$ according to (15.6) obtained by dividing by $t_n - t_{n-1}$.

15.4   Generalization

We shall also meet the following generalization of (15.2)

$$x'(t) = f(x(t), t) \tag{15.12}$$

in which the function f on the right hand side depends not only on t but also on the unknown solution $x(t)$. The analog of formula (15.3) now may take the form

$$x(t_n) \approx x(t_{n-1}) + f(x(t_{n-1}), t_{n-1})(t_n - t_{n-1}), \tag{15.13}$$

or changing from t_{n-1} to t_n in the f-term and recalling (15.4),

$$x(t_n) \approx x(t_{n-1}) + f(x(t_n), t_n)(t_n - t_{n-1}), \tag{15.14}$$

where as above, $0 = t_0 < t_1 < \cdots < t_{n-1} < t_n \cdots < t_N = 1$ is a sequence of time instances.

Using (15.13), we may successively determine approximations of $x(t_n)$ for $n = 1, 2, ..., N$, assuming that $x(t_0)$ is a given initial value. If we use instead (15.14), we obtain in each step an algebraic equation to determine $x(t_n)$ since the right hand side depends on $x(t_n)$.

In this way, solving the differential equation (15.12) approximately for $0 < t < 1$ is reduced to computing $x(t_n)$ for $n = 1, ..., N$, using the explicit formula (15.13) or solving the algebraic equation $X_n = X(t_{n-1}) + f(X_n, t_n)(t_n - t_{n-1})$ in the unknown X_n.

A basic example is the differential equation

$$x'(t) = x(t) \quad \text{for } t > 0, \tag{15.15}$$

corresponding to choosing $f(x(t), t) = x(t)$. In this case (15.13) takes the form

$$x(t_n) \approx x(t_{n-1}) + x(t_{n-1})(t_n - t_{n-1}) = (1 + (t_n - t_{n-1}))x(t_{n-1}).$$

With $(t_n - t_{n-1}) = \frac{1}{N}$ constant for $n = 1, ..., N$, we get the formula

$$x(t_n) \approx (1 + \frac{1}{N})x(t_{n-1}) \quad \text{for } n = 1, ..., N.$$

Repeating this formula, we get $x(t_n) \approx (1 + \frac{1}{N})(1 + \frac{1}{N})x(t_{n-2})$, and so on, which gives

$$x(1) \approx (1 + \frac{1}{N})^N x(0). \tag{15.16}$$

In fact, there is an exact solution of the equation $x'(t) = x(t)$ for $t > 0$ satisfying $x(0) = 1$, denoted by $x(t) = \exp(t)$, which is the *exponential function*. The formula (15.16) gives the following approximate formula for $\exp(1)$, where $\exp(1) = e$ is commonly referred to as the *base of the natural logarithm*:

$$e \approx (1 + \frac{1}{N})^N. \tag{15.17}$$

We give below values of $(1 + \frac{1}{N})^N$ for different N:

N	$(1+\frac{1}{N})^N$
1	2
2	2.25
3	2.37
4	2.4414
5	2.4883
6	2.5216
7	2.5465
10	2.5937
20	2.6533
100	2.7048
1000	2.7169
10000	2.7181

The differential equation $x'(t) = x(t)$ for $t > 0$, models the evolution of for example a population of bacteria which grows at a rate $x'(t)$ equal to the given amount of bacteria $x(t)$ at each time instant t. After each one time unit such a population has multiplied with the factor $e \approx 2.72$.

15.5 Leibniz' Teen-Age Dream

A form of Calculus was envisioned by Leibniz already as a teen-ager. Young Leibniz used to amuse himself with tables of the following form

n	1	2	3	4	5	6	7
n^2	1	4	9	16	25	36	49
	1	3	5	7	9	11	13
	1	2	2	2	2	2	2

or

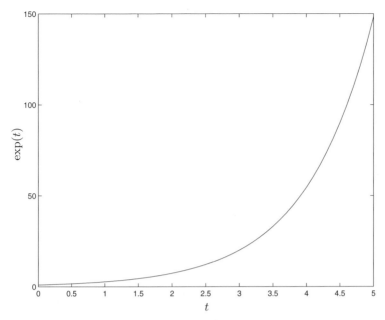

FIGURE 15.4. Graph of $\exp(t)$: Exponential growth.

n	1	2	3	4	5
n^3	1	8	27	64	125
	1	7	19	37	61
	1	6	12	18	24
	1	5	6	6	6

The pattern is that below each number, one puts the difference of that number and the number to its left. From this construction, it follows that any number in the table is equal to the sum of all the numbers in the next row below and to the left of the given number. For example for the squares n^2 in the first table we obtain the formula

$$n^2 = (2n - 1) + (2(n - 1) - 1) + \cdots + (2 \cdot 2 - 1) + (2 \cdot 1 - 1), \quad (15.18)$$

which can also be written as

$$n^2 + n = 2(n + (n - 1) + \cdots + 2 + 1) = 2 \sum_{k=1}^{n} k. \quad (15.19)$$

This corresponds to the area of the "triangular" domain in Fig. 15.5, where each term in the sum (the factor 2 included) corresponds to the area of one of the colons of squares.

The formula (15.19) is an analog of the formula

$$x^2 = 2 \int_0^x y \, dy$$

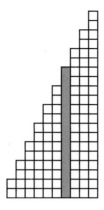

FIGURE 15.5.

with x corresponding to n, y to k, dy to 1, and $\sum_{k=1}^{n}$ to $\int_{0}^{n}$. Note that for n large the n-term in (15.19) is vanishing in comparison with n^2 in the sum $n^2 + n$.

By dividing by n^2, we can also write (15.18) as

$$1 = 2 \sum_{k=1}^{n} \frac{k}{n} \frac{1}{n} - \frac{1}{n}, \tag{15.20}$$

which is an analog of

$$1 = 2 \int_{0}^{1} y \, dy$$

with dy corresponding to $\frac{1}{n}$, y to $\frac{k}{n}$ and $\sum_{k=0}^{n}$ to $\int_{0}^{1}$. Note that the term $-\frac{1}{n}$ in (15.20) acts as a small error term that gets smaller with increasing n.

From the second table with n^3 we may similarly see that

$$n^3 = \sum_{k=1}^{n} (3k^2 - 3k + 1), \tag{15.21}$$

which is an analog of the formula

$$x^3 = \int_{0}^{x} 3y^2 \, dy$$

with x corresponding to n, y to k and $dy = 1$.

By dividing by n^3, we can also write (15.21) as

$$1 = \sum_{k=0}^{n} 3\left(\frac{k}{n}\right)^2 \frac{1}{n} - \frac{1}{n} \sum_{k=0}^{n} 3\frac{k}{n} \frac{1}{n} + \frac{1}{n^2},$$

which is an analog of

$$1 = \int_0^1 3y^2 \, dy$$

with dy corresponding to $\frac{1}{n}$, y to $\frac{k}{n}$, and $\sum_{k=0}^n$ to $\int_0^1$. Again, the error terms that appear get smaller with increasing n.

Notice that repeated use of summation allows e.g. n^3 to be computed starting with the constant differences 6 and building the table from below.

15.6   Summary

We may think of Calculus as the science of solving differential equations. With a similar sweeping statement, we may view Linear Algebra as the science of solving systems of algebraic equations. We may thus present the basic subjects of our study of Linear Algebra and Calculus in the form of the following two problems:

$$\text{Find } x \text{ such that } f(x) = 0 \quad \text{(algebraic equation)} \qquad (15.22)$$

where $f(x)$ is a given function of x, and

$$\text{Find } x(t) \text{ such that } x'(t) = f(x(t), t)$$
$$\text{for } t \in (0, 1], \ x(0) = 0, \quad \text{(differential equation)} \quad (15.23)$$

where $f(x, t)$ is a given function of x and t. Keeping this crude description in mind when following this book may help to organize the jungle of mathematical notation and techniques inherent to Linear Algebra and Calculus.

We take a *constructive* approach to the problem of solving equations, where we seek *algorithms* through which solutions may be determined or computed with more or less work. Algorithms are like recipes for finding solutions in a step by step manner. In the process of constructively solving equations one needs *numbers* of different kinds, such as *natural numbers, integers, rational numbers*. One also needs the concept of *real numbers, real variable, real-valued function, sequence of numbers, convergence, Cauchy sequence* and *Lipschitz continuous function*.

These concepts are supposed to be our humble servants and not terrorizing masters, as is often the case in mathematics education. To reach this position we will seek to de-mystify the concepts by using the constructive approach as much as possible. We will thus seek to look behind the curtain on the theater scene of mathematics, where often very impressive looking phenomena and tricks are presented by math teachers, and we will see that as students we can very well make these standard tricks ourselves, and in fact come up with some new tricks of our own which may even be better than the old ones.

15.7 Leibniz: Inventor of Calculus and Universal Genius

Gottfried Wilhelm von Leibniz (1646-1716) is maybe the most versatile scientist, mathematician and philosopher all times. Newton and Leibniz independently developed different formulations of Calculus; Leibniz notation and formalism quickly became popular and is the one used still today and which we will meet below.

FIGURE 15.6. Leibniz, Inventor of Calculus: "Theoria cum praxis". "When I set myself to reflect on the Union of Soul with the Body, I seemed to be cast back again into the open sea. For I could find no way of explaining how the Body causes something to happen in the Soul, or vice versa.....Thus there remains only my hypothesis, that is to say *the way of the pre-established harmony*–pre-established, that is by a Divine anticipatory artifice, which is so formed each of theses substances from the beginning, that in merely following its own laws, which it received with its being, it is yet in accord with the other, just as if they mutually influenced one another, or as if, over and above his general concource , God were for ever putting in his hands to set them right."

Leibniz boldly tackled the basic problem in Physics/Philosophy/Psychology of *Body and Soul* in his treatise *A New System of Nature and the Communication of Substances as well as the Union Existing between the Soul and the Body* from 1695. In this work Leibniz presented his theory of *Pre-established Harmony of Soul and Body*, In the related *Monadology* he describes the World as consisting of some kind of *elementary particles* in the form of *monads*, each of which with a blurred incomplete perception of the rest of the World and thus in possession of some kind of primitive soul. The modern variant of Monadology is *Quantum Mechanics*, one of the most spectacular scientific achievements of the 20th century.

Here is a description of Leibniz from Encyclopedia Britannica: "Leibniz was a man of medium height with a stoop, broad-shouldered but bandy-legged, as capable of thinking for several days sitting in the same chair as of travelling the roads of Europe summer and winter. He was an indefatigable worker, a universal letter writer (he had more than 600 correspondents), a patriot and cosmopolitan, a great scientist, and one of the most powerful spirits of Western civilization".

16
The Solar System

There is talk of a new astrologer who wants to prove that the earth moves and goes around instead of the sky, the sun, the moon, just as if somebody were moving in a carriage or ship might hold that he was sitting still and at rest while the earth and the trees walked and moved. But that is how things are nowadays: when a man wishes to be clever he must needs invent something special, and the way he does it must needs be the best! The fool wants to turn the whole art of astronomy upside-down. However, as Holy Scripture tells us, so did Joshua bid the sun to stand still and not the earth.
(Sixteenth century reformist M. Luther in his table book *Tischreden*, in response to Copernicus' pamphlet *Commentariolus*, 1514.)

16.1 Introduction

The problem of mathematical modeling of our solar system including the Sun, the nine planets Venus, Mercury, Tellus (the Earth), Mars, Jupiter, Saturn, Uranus, Neptune and Pluto together with a large number of moons and asteroids and occasional comets, has been of prime concern for humanity since the dawn of culture. The ultimate challenge concerns mathematical modeling of the Universe consisting of billions of galaxies each one consisting of billions of stars, one of them being our own Sun situated in the outskirts of the Milky Way galaxy.

According to the *geocentric* view presented by Aristotle (384-322 BC) in *The Heavens* and further developed by Ptolemy (87-150 AD) in *The Great*

System dominating the scene over 1800 years, the Earth is the center of the Universe with the Sun, the Moon, the other planets and the stars moving around the Earth in a complex pattern of circles upon circles (so-called epicycles). Copernicus (1473–1543) changed the view in *De Revolutionibus* and placed the Sun in the center in a new *heliocentric* theory, but kept the complex system of epicycles (now enlarged to a very complex system of 80 circles upon circles). Johannes Kepler (1572–1630) discovered, based on the extensive accurate observations made by the Swedish/Danish scientist Tycho Brahe (1546–1601), that the planets move in elliptic orbits with the Sun in one of the foci following *Kepler's laws*, which represented an enormous simplification and scientific rationalization as compared to the system of epicycles.

In fact, already Aristarchus (310-230 BC) of Samos understood that the Earth rotates around its axis and thus could explain the (apparent) motion of the stars, but these views were rejected by Aristotle arguing as follows: if the Earth is rotating, how is it that an object thrown upwards falls on the same place? How come this rotation does not generate a very strong wind? No one until Copernicus could question these arguments. Can you? *Yes.*

Newton (1642–1727) then cleaned up the theory by showing that the motion of the planets could be explained from one single hypothesis: the inverse square law of gravitation. In particular, Newton derived Kepler's laws for the *two-body problem* with one (small) planet in an elliptic orbit around a (large) sun. Leibniz criticized Newton for not giving any explanation of the inverse square law, which Leibniz believed could be derived from some basic fact, beyond one of "mutual love" which was quite popular. A sort of explanation was given by Einstein (1879–1955) in his theory of *General Relativity* with gravitation arising as a consequence of space-time being "curved" by the presence of mass. Einstein revolutionized *cosmology*, the theory of the Universe, but relativistic effects only add small corrections to Newton's model for our solar system based on the inverse square law. Einstein gave no explanation why space-time gets curved by mass, and still today there is no convincing theory of gravitation with its mystical feature of "action at a distance" through some mechanism yet to be discovered.

Despite the lack of a physical explanation of the inverse square law, Newton's theory gave an enormous boost to mathematical sciences and a corresponding kick to the egos of scientists: if the human mind was capable of (so easily and definitely) understanding the secrets of the solar system, then there could be no limits to the possibilities of scientific progress...

[Margin notes: "Simple geometry I think." "Mass is tightly curved space-time." "Lines of force spreading out in 3-D space ought to separate or spread out according to an inverse square law."]

16.2   Newton's Equation

The basis of celestial mechanics is Newton's second law,

$$F = m \cdot a, \qquad (16.1)$$

[Margin note: And here also, what are the units of measurment?]

FIGURE 16.1. Tycho Brahe: "I believe that the Sun and the Moon orbit around the Earth but that the other planets orbit around the Sun."

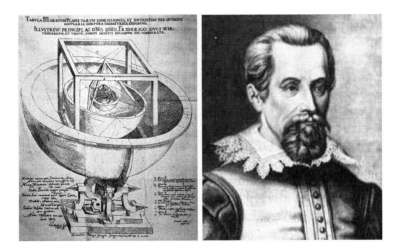

FIGURE 16.2. Johannes Kepler: "I believe that the planets are separated by invisible regular polyhedra: tetrahedron, cube, octahedron, dodekahedron and ikosahedron, and further that the planets including the Earth move in elliptical orbits around the Sun."

expressing that a *force* F results in an acceleration of size a for a body of mass m, together with the expression for the gravitational force given by the inverse square law:

$$F = G\frac{mm_a}{r^2},\qquad(16.2)$$

where $G \approx 6.67 \cdot 10^{-11}\mathrm{Nm^2/kg^2}$ is the *gravitational constant*, m_a is the mass of the attracting body and r is the distance to the attracting body.

Together (16.1) and (16.2) give a system of Ordinary Differential Equations (ODEs) for the evolution of the solar system, which take the form $\dot{u} = f(u)$, where the dot indicates differentiation with respect to time, $u(t)$ is a vector containing all the positions and velocities of the bodies and $f(u)$ is a given vector function of u. If we know the initial positions and velocities for all bodies in the solar system, we can find their positions and velocities at a later time by solving the system of differential equations $\dot{u} = f(u)$ by time-stepping. We discuss this in more detail below in Section 16.4. As a preparation, we rewrite (16.1) and (16.2) in dimensionless form, which will be convenient. The three fundamental units appearing in the equations are those of *space*, *time* and *mass*, which are represented by the variables x (or r), t and m. We now introduce new dimensionless variables, $x' = x/\mathrm{AU}$, $t' = t/\mathrm{year}$ and $m' = m/M$, where 1 AU is the mean distance from the Sun to Earth and M is the mass of the Sun. We can use the chain rule to obtain the dimensionless acceleration, $a' = \frac{d}{dt'}\frac{d}{dt'}x' = \frac{dt}{dt'}\frac{d}{dt}\frac{dt}{dt'}\frac{d}{dt}x/\mathrm{AU} = \frac{\mathrm{year}^2}{\mathrm{AU}}a$. Combining (16.1) and (16.2) using our new dimensionless variables, we then obtain

$$m'M\frac{\mathrm{AU}}{\mathrm{year}^2}a' = G\frac{m'M\cdot m_a'M}{r'^2\mathrm{AU}^2},\qquad(16.3)$$

or

$$a' = G'\frac{m_a'}{r'^2},\qquad(16.4)$$

where the new gravitational constant G' is given by

$$G' = \frac{G\cdot\mathrm{year}^2 M}{\mathrm{AU}^3}.\qquad(16.5)$$

We leave it as an exercise to show that with suitable definitions of the units year and AU, the new dimensionless gravitational constant G' is given by

$$G' = 4\pi^2.\qquad(16.6)$$

16.3   Einstein's Equation

In general relativity the basic concept is not *force*, as in Newtonian theory, but instead the *curvature* of space-time. Einstein explains the motion of the planets in our solar system in the following way: the planets move through space-time along straight lines, *geodesics*, which appear as circular (or elliptical) orbits only because space-time is curved by the large mass of the Sun. We shall now try to give an idea of how this works.

The curvature of space-time is given by its *metric*. A metric defines the distance between two nearby points in space-time. In Euclidean geometry that we have studied extensively in this book, the distance between two points $x = (x_1, x_2, x_3)$ and $y = (y_1, y_2, y_3)$ is given by the square root of the scalar product $dx \cdot dx$, where dx is the difference $dx = x - y$. With the notation $ds = |x - y|$ we thus have

$$ds = \sqrt{dx \cdot dx} = \left(\sum_{i=1}^{3} dx_i^2 \right)^{1/2}, \tag{16.7}$$

or

$$ds^2 = \sum_{i=1}^{3} dx_i^2. \tag{16.8}$$

In the notation of general relativity, the Euclidean metric is then given by the matrix (tensor)

$$g = \begin{bmatrix} 1 & 0 & 0 \\ 0 & 1 & 0 \\ 0 & 0 & 1 \end{bmatrix}, \tag{16.9}$$

as

— what's T?

$$ds^2 = dx^T g \, dx. \tag{16.10}$$

In space-time we include time t as a fourth coordinate and every event in space-time is given by a vector (t, x_1, x_2, x_3). In flat or *Minkowski* space-time in the absence of masses, the curvature is zero and the metric is given by

$$g = \begin{bmatrix} -1 & 0 & 0 & 0 \\ 0 & 1 & 0 & 0 \\ 0 & 0 & 1 & 0 \\ 0 & 0 & 0 & 1 \end{bmatrix}, \tag{16.11}$$

which gives

$$ds^2 = -dt^2 + dx_1^2 + dx_2^2 + dx_3^2. \tag{16.12}$$

In the presence of masses, we obtain a different metric which does not even have to be diagonal. *Oh?*

From the metric g one can find the straight lines of space-time, which give the orbits of the planets. The metric itself is determined by the distribution of mass in space-time, and is given by the solution of Einstein's equation,

$$R_{ij} - \frac{1}{2} R g_{ij} = 8\pi T_{ij}, \tag{16.13}$$

where (R_{ij}) is the so-called *Ricci-tensor*, R is the so-called *scalar curvature* and (T_{ij}) is the so-called *stress-energy tensor*. Now (R_{ij}) and R depend on derivatives of the metric $g = (g_{ij})$ so (16.13) is a partial differential equation for the metric g.

The solution for the orbits of the planets obtained from Einstein's equation are a little different than the solution obtained from (16.4) given by Newton. Although the difference is small, it has been verified in observations of the orbit of the planet Mercury which is the planet closest to the Sun. We will not include these "relativistic effects" in the next section where we move on to the computation of the evolution of the solar system.

16.4   The Solar System as a System of ODEs

Ordinary Differential Equation

Looks like the superscript i is an index.

We now rewrite the second-order system of ODEs given by (16.4) as a first order system of the form $\dot{u} = f(u)$. We start by introducing coordinates $x^i(t) = (x_1^i(t), x_2^i(t), x_3^i(t))$ for all bodies in the solar system, including the nine planets, then Sun and the Moon. This gives a total of $n = 9 + 2 = 11$ bodies and a total of $3n = 33$ coordinates. To rewrite the equations as the first-order system $\dot{u} = f$ we need to include also the velocities of all bodies, $\dot{x}^i(t) = (\dot{x}_1^i(t), \dot{x}_2^i(t), \dot{x}_3^i(t))$, giving a total of $N = 6n = 66$ coordinates. We collect all these coordinates in the vector $u(t)$ of length N in the following order: x_1, x_2, x_3 *are evidently the 3 space coordinates.*

dot x $\dot{x}$ evidently means velocity of x

$$u(t) = (x_1^1(t), x_2^1(t), x_3^1(t), \dots, x_1^n(t), x_2^n(t), x_3^n(t),$$
$$\dot{x}_1^1(t), \dot{x}_2^1(t), \dot{x}_3^1(t), \dots, \dot{x}_1^n(t), \dot{x}_2^n(t), \dot{x}_3^n(t)), \tag{16.14}$$

so that the first half of the vector $u(t)$ contains the positions of all bodies and the second half contains the corresponding velocities.

To obtain the differential equation for $u(t)$, we take the time-derivative and notice that the derivative of the first half of $u(t)$ is equal to the second half of $u(t)$:   *How do we notice this?*

$$\dot{u}_i(t) = u_{3n+i}(t), \quad i = 1, \dots, 3n, \tag{16.15}$$

Where'd 34 come from?

i.e. for $n = 11$ we have $\dot{u}_1(t) = \dot{x}_1^1(t) = u_{34}(t)$ and so on.

The derivative of the second half of $u(t)$ will contain the second derivatives of the positions, i.e. the accelerations, and these are given by (16.4). Now (16.4) is written as a scalar equation and we have to rewrite it in vector form. For every body in the solar system, we need to compute the contribution to the total force on the body by summing the contributions from all other bodies. Assuming that we work in dimensionless variables (but writing x instead of x', m_i instead of m_i' and so on for convenience) we then need to compute the sum:

Argh! Yet more shorthand, x for x'.

$$\ddot{x}_i(t) = \sum_{j \neq i} \frac{G'm_j}{|x^j - x^i|^2} \frac{x^j - x^i}{|x^j - x^i|}, \tag{16.16}$$

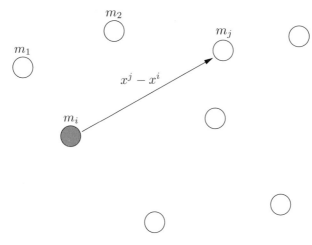

FIGURE 16.3. The total force on body i is the sum of the contributions from all other bodies.

where the unit vector $\frac{x^j - x^i}{|x^j - x^i|}$ gives the direction of the force, see Figure 16.3.

Our final differential equation for the evolution of the solar system in the form $\dot{u} = f$ is then given by

$$\dot{u}(t) = f(u(t)) = \begin{bmatrix} u_{3n+1}(t) \\ \vdots \\ u_{6n}(t) \\ \sum_{j \neq 1} \frac{G' m_j}{|x^j - x^1|^2} \frac{x_1^j - x_1^1}{|x^j - x^1|} \\ \vdots \\ \sum_{j \neq n} \frac{G' m_j}{|x^j - x^n|^2} \frac{x_3^j - x_3^n}{|x^j - x^n|} \end{bmatrix}, \qquad (16.17)$$

where we have kept the notation $x^1 = (x_1^1, x_2^1, x_3^1)$ rather than (u_1, u_2, u_3) and so on in the right-hand side for simplicity. The evolution of our solar system can now be computed using the techniques of time-stepping presented in e.g. Body&Soul. We may use the initial data supplied in Table 16.1. on the next page.

16.5   Predictability and Computability

Two important questions that arise naturally when we study numerical solutions of the evolution of our solar system, such as the one in Figure 16.4, are the questions of *predictability* and *computability*.

Are these exponents actually indices?

Nine planets

Sun

Moon

		Position	Velocity	Mass
Mercury	$x^1(0) =$	-0.147853935 -0.400627944 -0.198916163	$\dot{x}^1(0) =$ 7.733816715 -2.014137426 -1.877564183	$1.0/6023600$
Venus	$x^2(0) =$	-0.725771746 -0.039677000 0.027897127	$\dot{x}^2(0) =$ 0.189682646 -6.762413869 -3.054194695	$1.0/408523.5$
Earth	$x^3(0) =$	-0.175679599 0.886201933 0.384435698	$\dot{x}^3(0) =$ -6.292645274 -1.010423954 -0.438086386	$1.0/328900.5$
Mars	$x^4(0) =$	1.383219717 -0.008134314 -0.041033184	$\dot{x}^4(0) =$ 0.275092348 5.042903370 2.305658434	$1.0/3098710$
Jupiter	$x^5(0) =$	3.996313003 2.731004338 1.073280866	$\dot{x}^5(0) =$ -1.664796930 2.146870503 0.960782651	$1.0/1047.355$
Saturn	$x^6(0) =$	6.401404019 6.170259699 2.273032684	$\dot{x}^6(0) =$ -1.565320566 1.286649577 0.598747577	$1.0/3498.5$
Uranus	$x^7(0) =$	14.423408013 -12.510136707 -5.683124574	$\dot{x}^7(0) =$ 0.980209400 0.896663122 0.378850106	$1.0/22869$
Neptune	$x^8(0) =$	16.803677095 -22.983473914 -9.825609566	$\dot{x}^8(0) =$ 0.944045755 0.606863295 0.224889959	$1.0/19314$
Pluto	$x^9(0) =$	-9.884656563 -27.981265594 -5.753969974	$\dot{x}^9(0) =$ 1.108139341 -0.414389073 -0.463196118	$1.0/150000000$
Sun	$x^{10}(0) =$	-0.007141917 -0.002638933 -0.000919462	$\dot{x}^{10}(0) =$ 0.001962209 -0.002469700 -0.001108260	1
Moon	$x^{11}(0) =$	-0.177802714 0.884620944 0.384016593	$\dot{x}^{11}(0) =$ -6.164023246 -1.164502534 -0.506131880	$1.0/2.674 \cdot 10^7$

TABLE 16.1. Initial data for the solar system at 00.00 Universal Time (UT1, approximately GMT) January 1 2000 for dimensionless positions and velocities scaled with units 1 AU $= 1.49597870 \cdot 10^{11}$ m (one astronomical unit), 1 year $= 365.24$ days and $M = 1.989 \cdot 10^{30}$ kg (one solar mass).

23 Oct 2005

I guess m is meters & kg is kilograms

What frame of reference are these positions and velocities measured in relationship to? The "fixed" stars.

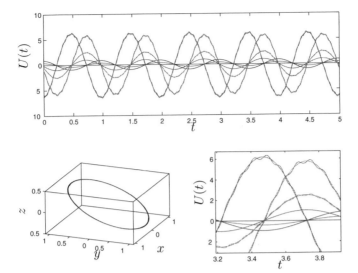

FIGURE 16.4. A numerical computation of the evolution of the solar system, including Earth, the Sun and the Moon.

The predictability of the solar system is the question of the accuracy of a computation given the accuracy in initial data. If initial data is known with an accuracy of say five digits, and the numerical computation is exact, how long does it take until the solution is no longer accurate even to one digit?

The computability of the solar system is the question of the accuracy in a numerical solution given exact initial data, i.e. how far we can compute an accurate solution with available resources such as method, computational power and time.

Both the predictability and the computability are determined by the rate errors grow. Luckily, errors do not grow exponentially in the case of the solar system. If we imagine that we displace Earth slightly from its orbit and start a computation, the orbit and velocity of Earth will be slightly different, resulting in an error that grows *linearly* with time. This means that the predictability of the solar system is quite good, since every extra digit of accuracy in initial data means that the limit of predictability is increased by a factor ten. If now the solution is computed using a computational time-stepping method, this will result in additional errors. We can think of the error from computation as a small perturbation introduced with every new time step. Adding the contributions from all time steps we may expect that the computational error typically grows *quadratically* in time. That is as x^2.

As it turns out however, for some computational methods the error grows only linearly: This is the case for the so called continuous Galerkin method with degree 1 piecewise polynomial approximation cG(1), as shown in Fig-

continuous Galerkin 1

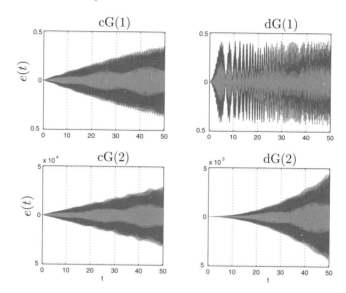

FIGURE 16.5. The growth of the numerical error in simulations of the solar system using different numerical methods. The two methods on the left conserve energy, which results in linear rather than quadratic error growth.

ure 16.5. This pleasant surprise is the result of an important property of the cG(1) method: it conserves energy. As a result, the cG(1) method performs better on a long time interval than the higher-order (more accurate) discontinuous Galerkin method dG(2) method.

What about using the rolling residual trick?
Does that have any relevance to cutting down error accumulati

16.6 Adaptive Time-Stepping

If we compute the evolution of the solar system using the adaptive cG(1) method, we find that the time steps need to be small enough to follow the orbit of the Moon (or Mercury if we do not include the Moon). This is inefficient since the time scales for the other bodies are much larger: the period of the Moon is one month and the period of Pluto is 250 years, and so *Oh?* the time steps for Pluto should be roughly a factor 3,000 larger that the time steps for the Moon. It has been shown recently that the standard methods cG(q), including cG(1), and dG(q) can be extended to individual, *multi-adaptive*, time-stepping for different components. In Figure 16.6 we show a computation made with individual time steps for the different planets. Notice how the error grows quadratically, indicating that the method does not conserve energy. (It is possible to construct also multi-adaptive methods which conserve energy.)

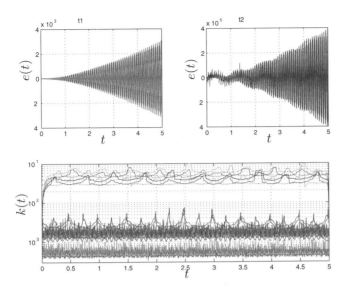

FIGURE 16.6. A computation of the evolution of the Solar System with individual, multi-adaptive, time steps for the different planets.

16.7   Limits of Computability and Predictability

Using the multi-adaptive cG(2) method, it appears that the limit of computability of the solar system (with the Moon and the nine planets) using double precision, is of the order 10^6 years. Concerning the predictability of the same system it appears that for every digit beyond 5 in the precision of data we gain a factor of ten in time, so that for example predicting the position of the Moon 1000 years ahead would require about 8 correct digits in e.g. the initial positions and velocities, masses and gravitational constant. We conclude that it appears that normally the precision in data would set the limit for accurate simulations of the evolution of the solar system, if we use a high order multi-adaptive solver.

17
Turbulence and the Clay Prize

I still remember his lectures. He was a mild-mannered, dapper man with a grey moustache, who squinted at his audience and lost it rather quickly. (Ivar Ekeland on Jean Leray 1906-1998)

Is it by accident that the deepest insight into turbulence came from Andrei Kolmogorov, a mathematician with a keen interest in the real world? (Uriel Frisch 1940-)

Since we don't even know whether these solutions exist, our understanding is at a very primitive level. Standard methods from PDE appear inadequate to settle the problem. Instead, we probably need some deep, new ideas. (offical Clay prize problem formulation)

A child, however, who had no important job and could only see things as his eyes showed them to him, went up to the carriage. "The Emperor is naked," he said. "Fool!" his father reprimanded, running after him. "Don't talk nonsense!" He grabbed his child and took him away. (HC Andersen 1805–1875)

17.1 The Clay Institute $1 Million Prize

We now return to the discussion of *turbulence* and the Navier–Stokes equations from the chapter *On the Reasonable Effectiveness of Computational Mathematics*. This connects to the seven *Clay Mathematics Institute* $1 million prize problems, one of which concerns the *existence*, *uniqueness* and *regularity* of solutions to the Navier–Stokes equations for incompressible fluid flow.

The Clay Institute prize problems were presented at the 2000 Millennium shift, as a reflection of the 23 problems formulated by the famous mathematician Hilbert at the second International Congress of Mathematicians in 1900 in Paris.

The prize problems are formulated to represent open problems of mathematics today of fundamental importance. By studying the formulations of these problems we may learn something about the dominating view within mathematics today.

We now focus on the formulation of the Navier–Stokes problem, which asks for a mathematical analytical proof of (i) existence and (ii) regularity or smoothness, as given by the famous mathematician Charles Fefferman (1949–), the winner of the Fields Medal 1978 (the Nobel Prize of mathematics). In this formulation of the problem, the uniqueness would follow from the regularity, and thus we may say that the uniqueness is also contained in the formulation of the problem. Regularity would mean that the solution could be differentiated many times, even if the derivatives become very large.

As a starting point one may take the only result available in this direction, which is a proof by the mathematician Jean Leray from 1934 of *existence* of so-called *weak solutions*, or *turbulent solutions* in the terminology used by Leray, where a weak solution satisfies the Navier–Stokes equations in an average sense. Thus, proving that weak solutions are unique and smooth, would give the $1 million prize. But nobody has been able to come up with such a proof, and the progress since Leray seems to be very small.

FIGURE 17.1. Claude Louis Marie Henri Navier (1785–1836), George Gabriel Stokes (1819–1903), Siméon Denis Poisson (1781–1840), and Adhémar Jean Claude Barré de Saint-Venant (1797–1886), who discovered the Navier–Stokes equations 1821–45.

What is now very intriguing is that the prize problem formulation seems to be based on a confusion. At least this is the impression one may get after a study of the problem using computational mathematics, or maybe it is our own view which is confused? To try to come to a conclusion, let's take a closer look at the formulation of the Clay prize problem.

17.2   Are Turbulent Solutions Unique and Smooth?

We know that Leray in 1934 proved the existence of what he called a weak solution or a turbulent solution. Of course, Leray was well aware of observations of turbulent flow in nature, with a turbulent flow being quickly fluctuating in space and time in a seemingly chaotic way. Leray knew from the experiments of the famous scientist Osborne Reynolds from the 1880s that turbulent flow occurs if the so called *Reynolds number* Re of the flow is large enough, where $Re = \frac{UL}{\nu}$ with U a characteristic flow velocity, L a characteristic length scale, and $\nu > 0$ the *viscosity* of the fluid. If Re is relatively small ($Re \leq 10 - 100$), then the flow is viscous and the flow field is ordered and smooth or *laminar*, while for larger Re, the flow will at least partly be *turbulent* with time-dependent non-ordered features on a range of length scales down to a smallest scale (which may be estimated to be of size $Re^{-3/4}$, assuming $L = 1$). If $Re = 10^6$, which is common in industrial applications, then the smallest vortex of the flow could be of size $10^{-4} - 10^{-5}$ if the length scale $L = 1$, thus of the size of a fraction of a millimeter if the overall diameter of the fluid volume is one meter. A turbulent flow thus could show a very complex interaction of vortices on a large range of different scales.

[margin handwriting: length scale et Hiesenberg indeterminancy?]

The Reynolds number of the flow of air around our car travelling at 90 km/h (≈ 60 mph) is about 10^6, and the flow of air is partly turbulent, in particular in the large *wake* attaching to the rear of the car, where the air is rapidly fluctuating and is seemingly chaotic. We can easily observe this wake on the highway in the case of light rain or mist when the flow pattern becomes visible.

We can also observe turbulent flow, for example, in a river in the wake behind a stone, which was systematically studied already by Leonardo da Vinci (1452–1519). Although Leonardo did not have access to calculus, he skillfully captured many essential features of turbulent flow in his sketches.

We may expect a laminar flow to be determined pointwise in space-time, while in a turbulent flow, because of its rapid fluctuations, we can only expect various mean values to be uniquely determined. In many applications of scientific and industrial importance Re is very large, of the order 10^6 or larger, and the flow shows a combination of laminar and turbulent features.

[margin handwriting: point-wise?]

Now, believing that the Navier–Stokes equations describes nature, Leray of course expected the Navier–Stokes equations to have turbulent solutions in the case of large Reynolds numbers, and with this perspective it was natural for him to call his weak solutions also turbulent solutions. But Leray did not prove uniqueness, and presumably he did not even try, because of the seemingly chaotic nature of a turbulent flow, uniqueness would seem to be out of the question. At least if we talk about uniqueness in the pointwise sense in space and time: since the turbulent solution is rapidly fluctuating in a seemingly chaotic fashion, it would be impossible to speak about (or measure) the exact value of the velocity of fluid particles at a specific point in space and time. But of course Leray may have expected that certain mean values in space and time could be more well determined (or uniquely determined), but he could not prove any such result.

FIGURE 17.2. Left: Jean Leray (1906–98) proved existence of weak solutions, Jacques Salomon Hadamard (1865–1963) first studied well-posedness of differential equations. Right: Leonardo da Vinci (1452–1519) sketch of turbulent wakes.

17.3   Well-Posedness According to Hadamard

The general question of uniqueness directly couples to a question about *well-posedness* of a set of differential equations, as first studied by the French mathematician Jacques Salomon Hadamard (1865–1963). A set of partial differential equations (like the Navier–Stokes equations) would be *well-posed* if small variations in data (like initial data) would result in small variations of the solution (at a later time). Hadamard stated that only well-posed mathematical models could be meaningful: if very small changes in data could cause large changes in the solution, it would clearly be impossible to reach the basic requirement in science of reproducible results.

The question of well-posedness may alternatively be viewed as a question of *sensitivity to perturbations*. A problem with very strong sensitivity to perturbations would not be well-posed in the Hadamard sense. Now Hadamard proved the well-posedness of some basic partial differential equations like the Poisson equation, but he did not state any result for the Navier–Stokes equations.

Of course, believing that solutions to the Navier–Stokes equations may be turbulent, and observing the seemingly chaotic nature of turbulence, we could not expect the Navier–Stokes equations to be well-posed in a pointwise sense: we would expect to see a very strong pointwise sensitivity to small perturbations. But, of course it would be natural to ask if certain mean values may be less sensitive, so that the Navier–Stokes equations would be well-posed in the sense of such mean values.

Again, mean ≡ average.

17.4   Is Mathematics a Science?

We now return to the formulation of the Clay prize problem. Since the formulation asks for uniqueness and smoothness, it would seem that possible turbulent solutions are not considered, because turbulent solutions cannot be expected to be neither pointwise unique in space-time, nor smooth since they are so rapidly fluctuating. It would rather seem as if only laminar solutions are considered. Thus, this formulation would make sense in the case of relatively small Reynolds numbers, but not in the case of large Reynolds numbers.

Now, how could it be possible to make such an elementary mistake? Or is not a mistake? Is it so that one may insist that also a turbulent solution, *in principle*, would be pointwise uniquely determined and also *in principle* smooth, if yet with very large derivatives? Yes, it appears that this could be the view underlying the formulation of the Clay prize problem. In particular, it appears that insisting on the uniqueness in principle, could allow a very strong sensitivity to perturbations.

In fact, it is easy to come to this conclusion by the fact that the uniqueness is not explicitly included in the formulation of the problem. This is presumably due to the fact that it is very easy to formally obtain uniqueness if it is known that the first derivatives of the velocity are bounded: if the bound is K one obtains, using a simple so-called Grönwall estimate, that the effect of an initial perturbation of size δ after a time T could be $e^{KT}\delta$. If $Re = 10^6$, then we expect that $K = 10^3$, and thus with $T = 10$ the amplification of the initial perturbation would be a factor e^{10000} which is much larger than 10^{100}, which is an incredibly large number, also referred to as a *googol*. But with such a strong sensitivity, anything could result from virtually nothing, and the scientific meaning would seem to be lost.

So, we come to the conclusion that the formulation of the Clay prize problem would insist that also a turbulent solution could be viewed as a unique and smooth solution, more precisely, as a solution with very large derivatives and an incredibly strong pointwise sensitivity to perturbations. But such a view on uniqueness would seem to be against the idea of well-posedness by Hadamard as a requirement of science. When we discuss this point with a leading mathematical expert on the Navier–Stokes equations, supporting the view of the Clay problem formulation, we get the message that science is one thing, and mathematics something different: in science an amplification factor of a googol would be way too large for any kind of uniqueness. But it appears that in mathematics it would be acceptable in a proof of uniqueness. But is this point of view reasonable and constructive, or the opposite?

We now address the question of uniqueness using computational mathematics. It is then natural to formulate the basic question in quantitative terms as follows for a given flow situation: What quantity of interest or output can be computed to what tolerance to what cost?

We consider the problem of computing the *drag* of a *bluff body*, which could be the force our car meets from the surrounding air at a speed of 90 km/h ($\approx$ 60 mph), which directly would couple to the fuel consumption. We could then seek to compute the drag $D(t)$ at a specific time t, or the mean value of $D(t)$ over some time interval. The c_D-coefficient of our car is such a mean value over very long time. So our quantities of interest or outputs may be the momentary value of the drag $D(t)$ at a given specific time t and the mean value of $D(t)$ over different time intervals.

We show that we may reliably compute the long-time mean value c_D on a PC within hours up to a tolerance of a few percent. We also demonstrate an increasing difficulty of computing $D(t)$ over shorter time intervals, and in particular that the momentary value at a given time instant appears to be uncomputable, because $D(t)$ is rapidly fluctuating in time with a total variation of about 30%. Moreover, each record of the variation in time of $D(t)$ seems different, and it thus appears impossible to assign a specific value of $D(t)$ to a particular time instant t. Just as it appears impossible to assign a specific weather and temperature to London at July 1st each year. While to speak of a typical mean value temperature over the month of July, would seem perfectly possible. In fact, guide books often supply such data, while they (wisely) avoid to give predictions of specific temperatures on e.g. July 1st.

We could then phrase these results in terms of output uniqueness/non-uniqueness of weak solutions. Of course, if we consider pointwise outputs also in space (the drag $D(t)$ is a mean value in space of pointwise forces), then the non-uniqueness may be expected to become even more pronounced.

The reader who does not want details could stop here, with the main conclusions from our computations being made: certain mean value outputs are unique/computable, while pointwise values in space-time are not.

For the reader who wants more substance, we now present some more detailed information on the computations.

17.5   A Computational Approach to the Problem

Let us now follow a line of thought that Leray may have taken today, with computational methods for solving the Navier–Stokes equations available. The weak solution of Leray would then be a computed solution obtained by using a computational method such as the *finite element method*. In fact, using the finite element method we seek a kind of approximate weak solution in the form of a piecewise polynomial (e.g. piecewise linear function) on a subdivision (mesh) of space-time. Since we can compute finite element solutions, we are not surprised to see the existence of approximate solutions, and the pertinent question then becomes uniqueness. Since the computed solution is (partly) turbulent it will look chaotic, and thus we cannot expect to see pointwise convergence as we refine the mesh, and in fact in computations we don't see it. The pertinent question then is the same: what mean values in space-time can we compute reliably?

Using a computational approach we may study the question of existence and uniqueness for a set of specific cases with given data, which may be representative for a wider selection of data, but we will not be able to give one answer for all possible data, as the ideal analytical mathematical proof would give. We are thus restricted to a case by case study, and a reformulation of the prize problem as a set of 10^3 prizes each of 10^3, would seem more natural.

Below we shall give also computational evidence of the existence of turbulent solutions. So even if we cannot analytically construct turbulent solutions to the Navier–Stokes equations, we can observe turbulent flow in real life and we can also compute approximate solutions which are turbulent. Of course it is natural to expect that the computed solutions approximate the weak (turbulent) solutions proved to exist by Leray.

We now focus on flows at moderate to large Reynolds numbers, where we thus expect to meet both laminar and turbulent flow features. Normalizing the flow velocity U and the length scale L both to one, we thus focus on flows with *small viscosity* ν, say typically $\nu \leq 10^{-6}$.

As already indicated, in the case of turbulent flow it is natural to seek to compute, instead of pointwise quantities some more or less local *mean values* in space-time. More precisely, we choose as a *quantity of interest* to compute or *output*, a certain mean value. In the case of the car it may be a mean value in time of the total *drag force* $D(t)$ at time t acting on the car in the direction opposite to the motion of the car. The consumption of fuel of a car is directly related to the mean value in time of the drag force $D(t)$, which suitably normalized is referred to as the c_D-coefficient, or *drag coefficient*. Some car manufacturers like to present the c_D of a certain car as an indication of fuel economy (for example $c_D < 0.3$). For a jumbo-jet a decrease in drag with one percent could save $400 million in fuel cost over a 25 year life span.   *Eee gads!*

So we may ask, for example, if the c_D of a car would be uniquely determined? Or in the setting of weak solutions: Will two weak solutions give the same c_D? The corresponding normalized mean value in time of the total force perpendicular to the direction of motion is referred as the *lift coefficient* c_L, which is crucial for flying vehicles (or sailing boats and also very fast cars).   *c_L denotes coefficient of lift.*

We will approach this type of problem by computational methods, and it is then natural to rephrase the problem as a problem of *computability*. We then specify an output, an *error tolerance* TOL, a certain amount of computational work W (or computational cost), and we ask if we can compute the output up to the tolerance TOL with the available work W.

How about interval quantities?

For example, we may ask if we can compute the c_D coefficient of a specific car up to a tolerance of 5% on our PC within 1 hour?

More generally we propose the following formulation of the Clay prize problem:

- (PC) For a given flow, what output can be computed to what tolerance to what cost?

We may view (PC) as a computational version of the question of uniqueness of a weak solution of the form:

What the heck is PWO?

- (PWO) Is the output of a weak solution unique?

(PWO) can alternatively be phrased as a question of well-posedness in a weak sense. We refer to (PWO) as a question of *weak uniqueness* with respect to a given output. Below we will approach the questions of weak uniqueness of the mean value c_D and the momentary value $D(t)$ of the drag force. Of course, (PWO) couples to the concept *observable quantities* of basic relevance in physic. It may seem that only uniqueness of observable quantities could be the subject of scientific investigation. This couples to questions of classical vs quantum mechanics, e.g. the question if an electron can be located at a specific point in time and space.

We will now address (PC) using the technique of adaptive finite element methods with a posteriori error estimation. The a posteriori error estimate results from an *error representation* expressing the output error as a space-time integral of the *residual* of a computed solution multiplied by *weights* which relate to derivatives of the solution of an associated *dual problem*. The weights express *sensitivity* of a certain output with respect to the residual of a computed solution, and their size determine the degree of computability of a certain output: The larger the weights are, the smaller the residual has to be and the more work is required. In general the weights increase as the size of the mean value in the output decreases, indicating increasing computational cost for more local quantities. We give computational evidence in a bluff body problem that a mean value in time of the drag (the c_D) is computable to a reasonable tolerance at a reasonable computational cost, while the value of the drag at a specific point in time appears to be uncomputable even at a very high computational cost.

We can rephrase this result for (PC) as the following result for (PWO): Two weak solutions of a bluff body problem give the same c_D. At least we have then given computational evidence of a certain output uniqueness of weak solutions.

As a general remark on approximate solutions obtained using the finite element method, we recall that a finite element solution is set up to be an approximate weak solution, and thus there is a strong connection between finite element solutions and weak solutions.

Leray's proof of existence of weak solutions is based on a basic energy estimate for approximate solutions of the Navier–Stokes equations, which

And here we are getting close to Heisenberg principle of indeterminacy

could be finite element solutions. Using the basic energy estimate one may extract a weakly convergent subsequence of approximate solutions as the mesh size tends to zero, and this way obtain a proof of existence of a weak solution. Even if the finite element solution on each given mesh is unique, a weak limit of a sequence of finite element solutions does not have to be unique, and thus the Leray solution is not necessarily unique. Of course, with this perspective the questions (PWO) and (PC) become closely coupled: (PWO) is close to the question of output uniqueness of a weak limit of a sequence of finite element solutions, which is close to the output computability (PC).

17.6   The Navier–Stokes Equations

The Navier–Stokes equations for an incompressible fluid with constant kinematic viscosity $\nu > 0$ occupying a volume Ω in $\mathbb{R}^3$ with boundary Γ, take the form: *Oy Veh! Haven't the slightest idea what ∇ & Δ are, except probably some sort of vector functions.*

$$
\begin{aligned}
\dot{u} + (u \cdot \nabla)u - \nu \Delta u + \nabla p &= f && \text{in } \Omega \times I, \\
\nabla \cdot u &= 0 && \text{in } \Omega \times I, \\
u &= 0 && \text{on } \Gamma \times I, \\
u(\cdot, 0) &= u^0 && \text{in } \Omega,
\end{aligned}
\tag{17.1}
$$

where $u(x,t) = (u_1(x,t), u_2(x,t), u_3(x,t))$ is the *velocity* and $p(x,t)$ the *pressure* of the fluid at $(x,t) = (x_1, x_2, x_3, t)$, and $f(x,t)$, $u^0(x)$, $I = (0,T)$, is a given driving force, initial data and time interval, respectively. For simplicity and definiteness we assume homogeneous Dirichlet boundary conditions for the velocity.   *No kidding*

The first equation in (17.1) expresses conservation of momentum (Newton's Second Law) and the second equation expresses conservation of mass in the form of incompressibility.

The Navier–Stokes equations formulated in 1821–45 appear to give an accurate description of fluid flow including both laminar and turbulent flow features. *Computational Fluid Dynamics CFD* concerns the computational simulation of fluid flow by solving the Navier–Stokes equations numerically. To computationally resolve all the features of a flow in a *Direct Numerical Simulation DNS* seems to require of the order Re^3 mesh points in space-time, so already a flow at $Re = 10^6$ would require $Re^3 = 10^{18}$ mesh points in space-time, and thus would seem to be impossible to solve on any foreseeable computer.

The computational challenge is to compute high Reynolds number flows (e.g $Re = 10^6$) using less computational effort than in a DNS. We shall see that for certain mean value outputs such as the c_D or c_L coefficients, this indeed appears to be possible: We give evidence that the c_D and c_L of a surface mounted cube may be computed on a PC up to a tolerance of a few percent (but not less).

17.7 The Basic Energy Estimate for the Navier–Stokes Equations

We now derive a basic stability estimate of energy type for the velocity u of the Navier–Stokes equations (17.1), assuming for simplicity that $f = 0$. This is about the only analytical a priori estimate known for the Navier–Stokes equations.

Scalar multiplication of the momentum equation by u and integration with respect to x gives

$$\frac{1}{2}\frac{d}{dt}\int_\Omega |u|^2\,dx + \nu\sum_{i=1}^3 \int_\Omega |\nabla u_i|^2\,dx = 0,$$

because by partial integration (with boundary terms vanishing),

$$\int_\Omega \nabla p \cdot u\,dx = -\int_\Omega p\nabla \cdot u\,dx = 0$$

and

$$\int_\Omega (u\cdot\nabla)u\cdot u\,dx = -\int_\Omega (u\cdot\nabla)u\cdot u\,dx - \int_\Omega \nabla\cdot u|u|^2\,dx$$

so that

$$\int_\Omega (u\cdot\nabla)u\cdot u\,dx = 0.$$

Integrating next with respect to time, we obtain the following basic a priori stability estimate for $T > 0$:

$$\|u(\cdot,T)\|^2 + D_\nu(u,T) = \|u^0\|^2,$$

$$D_\nu(u,T) = \nu\sum_{i=1}^3 \int_0^T \|\nabla u_i\|^2\,dt, \qquad (17.2)$$

Sure it does.

where $\|\cdot\|$ denotes the $L_2(\Omega)$-norm. This estimate gives a bound on the kinetic energy of the velocity with $D_\nu(u,T)$ representing the total *dissipation* from the viscosity of the fluid over the time interval $[0,T]$. We see that the growth of this term with time corresponds to a decrease of the velocity (momentum) of the flow (with $f = 0$).

The characteristic feature of a turbulent flow is that $D_\nu(u,T)$ is comparatively large, while in a laminar flow with ν small, $D_\nu(u,T)$ is small. With $D_\nu(u,T) \sim 1$ in a turbulent flow and $|\nabla u|$ uniformly distributed, we may expect to have pointwise

$$|\nabla u_i| \sim \nu^{-1/2}. \qquad (17.3)$$

17.8   Weak Solutions

From the basic energy estimate, Leray derived the existence of a weak solution $(u, p) \in V \times Q$ of the Navier–Stokes equations, defined by:

$$R_\nu(u, p; v, q) \equiv ((\dot{u}, v)) + ((u \cdot \nabla u, v)) - ((\nabla \cdot v, p)) + ((\nabla \cdot u, q))$$
$$+ ((\nu \nabla u, \nabla v)) - ((f, v)) = 0 \quad \forall (v, q) \in V \times Q, \tag{17.4}$$

assuming $u(0) = u^0 \in L_2(\Omega)^3$ and $f \in L_2(I; H^{-1}(\Omega)^3)$, where

$$V = \{v : v \in L_2(I; H_0^1(\Omega)^3), \dot{v} \in L_2(I; H^{-1}(\Omega)^3)\},$$
$$Q = L_2(I; L_2(\Omega)),$$

where $H_0^1(\Omega)^3$ is the usual Sobolev space of vector functions being square integrable together with their first derivatives over Ω, with dual $H^{-1}(\Omega)^3$, and $((\cdot, \cdot))$ denoting the corresponding $L_2(I; L_2(\Omega))$ inner product or pairing. As usual, $L_2(I; X)$ with X a Hilbert space denotes the set of functions $v : I \rightarrow X$ which are square integrable. Below we write $L_2(X)$ instead of $L_2(I; X)$ and $L_2(H^1)$ and $L_2(H^{-1})$ instead of $L_2(H_0^1(\Omega)^3)$ and $L_2(I; H^{-1}(\Omega)^3)$. Note that the term $((u \cdot \nabla u, v))$ is interpreted as

Yeah, the usual Sobolev space.

We all know what those are.

As usual. Uh huh...

$$-\sum_{i,j} ((u_i u_j, v_{j,i})),$$

where $v_{j,i} = \partial v_j / \partial x_i$.

17.9   Computational Solution

We now consider a computational solution of the Navier–Stokes equations. Without going into details of the construction of these methods, which we refer to as Generalized Galerkin or G^2, we can describe these methods as producing an approximate solution $(u_h, p_h) \in V_h \times Q_h$, where $V_h \times Q_h$ is a piecewise polynomial finite element subspace of $V \times Q$ defined on space-time meshes with h representing the mesh size in space-time, defined by the following discrete analog of (17.4)

I assume this h has nothing directly to do with Planck's constant.

$$R_h(u_h, p_h; v, q) = 0 \quad \text{for all } (v, q) \in V_h \times Q_h, \tag{17.5}$$

expressing that the discrete residual $R_h(u_h, p_h) = R_h(u_h, p_h : \cdot, \cdot)$ is orthogonal to $V_h \times Q_h$. Note that in the finite element method (17.5) we use an *artificial viscosity* of size h instead of the physical viscosity ν assuming $h > \nu$. There are other more sophisticated ways of introducing a (necessary) artificial viscosity coupled to weighted least squares stabilization in G^2, but here we consider the simplest form of stabilization.

The finite element solution satisfies an energy estimate analogous to (17.2) of the form

$$\|\sqrt{h}\nabla u_h\| \leq C, \tag{17.6}$$

where $\|\cdot\|$ denotes the $L_2(L_2)$-norm, which follows by choosing $(v, q) = (u_h, p_h)$ in (17.5). Here and below, C is a positive constant of unit size.

We will see below that to estimate an output error, we will have to estimate $R_\nu(u_h, p_h; \varphi_h, \theta_h)$, where (φ_h, θ_h) is the solution of a certain linear dual problem with data connected to the output. In general, (φ_h, θ_h) will not belong to the finite element subspace, and we will thus need to estimate $R_\nu(u_h, p_h; \varphi_h, \theta_h)$. The basic estimate for this quantity takes the form

$$|R_\nu(u_h, p_h; \varphi_h, \theta_h)| \leq C\sqrt{h}\|\varphi_h\|_{L_2(H^1)}, \tag{17.7}$$

if we omit the relevant θ_h-term assuming exact incompressibility, and C denotes a constant of moderate size. To motivate this estimate, we observe that estimating separately the dissipative term in G^2 (with viscosity h) using the energy estimate (17.6), we get by Cauchy's inequality

$$|((h\nabla u_h, \nabla \varphi_h))| \leq \|\sqrt{h}\nabla u_h\|\|\sqrt{h}\nabla \varphi_h\| \leq C\sqrt{h}\|\varphi_h\|_{L_2(H^1)}.$$

One can now argue that the remaining part of the residual can be estimated similarly, which leads to (17.7). We conclude that we expect the residual of (u_h, p_h) to be small (of size $h^{1/2}$) in a weak norm. However, we cannot expect the residual to be small in a strong sense: We would except the residual in an L_2-sense to be of size $h^{-1/2}$ reflecting the basic energy estimate (17.6), which suggests that $|\nabla u_h| \sim h^{-1/2}$ paralleling (17.3).

17.10   Output Error Representation

We now proceed to estimate the error in certain mean value outputs of a computed finite element solution (u_h, p_h) as compared to the output of a weak solution (u, p). We then consider an output of the form

$$M(u) = ((u, \psi))$$

where $\psi \in L_2(L_2)$ is a given (smooth) function. The output $M(u)$ then corresponds to a mean value in space and time of the velocity u with the function ψ appearing as a weight. We then establish an *error representation* in terms of the residual of the computed solution and the solution (φ_h, θ_h) of a certain linear dual problem (with coefficients depending on both u and u_h) to be specified below, of the form

$$M(u) - M(u_h) = R_\nu(u_h, p_h; \varphi_h, \theta_h). \tag{17.8}$$

We can then attempt to estimate the output error by using (17.7) to get

$$|M(u) - M(u_h)| \le C\sqrt{h}\|\varphi_h\|_{L_2(H^1)}, \qquad (17.9)$$

and the crucial question will thus concern the size of $\|\varphi_h\|_{L_2(H^1)}$. More precisely, we compute (an approximation of) the dual solution (φ_h, θ_h) and directly evaluate $R_\nu(u_h, p_h; \varphi_h, \theta_h)$. We may also use (17.9) to get a rough idea on the dependence of $M(u) - M(u_h)$ on the mesh size h, and we will then obtain convergence in output if, roughly speaking, $\|\varphi_h\|_{L_2(H^1)}$ grows slower than $h^{-1/2}$.

We need here to make the role of h vs ν more precise. We assume that ν is quite small, say $\nu \le 10^{-6}$, so that it is inconceivable that in a computation we could reach $h \le \nu$; we would rather have $10^{-4} \le h \le 10^{-2}$. In the finite element method we use an artificial viscosity of size h instead of the physical viscosity ν and thus computing on a sequence of meshes with decreasing h, could be seen as computing a sequence of solutions to problems with decreasing effective viscosity of size h. We would then be interested in the "limit" with $h = \nu$, and we would by observing the convergence (or divergence) for $h > \nu$ seek to draw a conclusion concerning the case $h = \nu$. So, in the computational examples to be presented we compute on a sequence of successively refined meshes with decreasing h and we evaluate the quantity $R_\nu(u_h, p_h; \varphi_h, \theta_h)$ to seek to determine convergence (or divergence) for a specific output.

17.11   The Dual Problem

The dual problem takes the following form, starting from a finite element solution (u_h, p_h) and a weak solution (u, p) with ψ a given (smooth) function: Find (φ_h, θ_h) with $\varphi_h = 0$ on Γ, such that

$$
\begin{aligned}
-\dot{\varphi}_h - (u \cdot \nabla)\varphi_h + \nabla u_h \cdot \varphi_h - \nu\Delta\varphi_h + \nabla\theta_h &= \psi & \text{in } \Omega \times I, \\
\nabla \cdot \varphi_h &= 0 & \text{in } \Omega \times I, \\
\varphi_h(\cdot, T) &= 0 & \text{in } \Omega,
\end{aligned}
$$
$$(17.10)$$

where $(\nabla u_h \cdot \varphi_h)_j = (u_h)_{,j} \cdot \varphi_h$. This is a linear convection-diffusion-reaction problem, where the time variable runs "backwards" in time with initial value $(= 0)$ given at final time T. The reaction coefficient ∇u_h is large and highly fluctuating, and the convection velocity u is of unit size and is also fluctuating. A standard Grönwall type estimate of the solution (φ_h, θ_h) in terms of the data ψ would bring in an exponential factor e^{KT} with K a pointwise bound of $|\nabla u_h|$ which would be enormous, as indicated above. When we compute the solution (φ_h, θ_h), corresponding to c_D or c_L, we note that (φ_h, θ_h) does not seem to explode exponentially at all, as would be indicated by Grönwall. Intuitively, by cancellation in the reaction term, with roughly as much production as consumption, (φ_h, θ_h) grows very slowly with deceasing h, and as we have said, the crucial question will be the growth of the quantity $\|\varphi_h\|_{L_2(H^1)}$.

To establish the error representation (17.8) we multiply (17.10) by $u - u_h$, integrate by parts, and use the fact that

$$(u \cdot \nabla)u - (u_h \cdot \nabla)u_h = (u \cdot \nabla)e + \nabla u_h \cdot e,$$

where $e = u - u_h$.

In the computations, we have to replace the convection velocity u by the computed velocity u_h. We don't expect u_h to necessarily be close pointwise to u, so we have to deal with the effect of a large perturbation in the dual linear problem. In the computations we get evidence that the effect on a crucial quantity like $\|\varphi_h\|_{L_2(H^1)}$ may be rather small, if the output is c_D or c_L. More precisely, our computations show in these cases a quite slow logarithmic growth of $\|\varphi_h\|_{L_2(H^1)}$ in terms of $1/h$, which indicates that the large perturbation in u indeed has little influence on the error representation for c_D and c_L.

The net result is that we get evidence of output uniqueness of weak solutions in the case the output is c_D or c_L. We contrast this with computational evidence that an output of the momentary drag $D(t)$ for a given specific point in time t, is not uniquely determined by a weak solution.

17.12   Output Uniqueness of Weak Solutions

Suppose we have two weak solutions (u, p) and $(\hat{u}, \hat{p})$ of the Navier–Stokes equations with the same data. Let (φ_h, θ_h) be a corresponding dual solution defined by the dual equation (17.10), with u_h replaced by $\hat{u}$, and a given output (given by the function ψ). Output uniqueness will then hold if $\|\varphi_h\|_{L_2(H^1)} < \infty$.

In practice, we will seek to compute $\|\varphi_h\|_{L_2(H^1)}$ approximately, replacing both u and $\hat{u}$ as coefficients in the dual problem by a computed solution u_h, thus obtaining an approximate dual velocity φ_h. We then study $\|\varphi_h\|_{L_2(H^1)}$ as h decreases and we extrapolate to $h = \nu$. If the extrapolated value $\|\varphi_\nu\|_{L_2(H^1)} < \infty$, or rather is not too large, then we have evidence of output uniqueness. If the extrapolated value is very large, we get indication of output non-uniqueness. As a crude test of largeness of $\|\varphi_\nu\|_{L_2(H^1)}$, it appears natural to use $\|\varphi_\nu\|_{L_2(H^1)} >> \nu^{-1/2}$.

We may further use a slow growth of $\|\varphi_h\|_{L_2(H^1)}$ as evidence that it is possible to replace both u and $\hat{u}$ by u_h in the computation of the solution of the dual problem: a near constancy indicates a desired robustness to (possibly large) perturbations of the coefficients u and $\hat{u}$.

We now proceed to give computational evidence.

17.13   Computational Results

Coefficient of drag [handwritten annotation]

Uniqueness of c_D and c_L *Coefficient of Lift?* [handwritten annotation]

The computational example is a bluff body benchmark problem. *bluff or rough?* [handwritten annotation] We compute the mean value in time of drag and lift forces on a surface mounted cube in a rectangular channel from an incompressible fluid governed by the Navier–Stokes equations (17.1), at $Re = 40.000$ based on the cube side length and the bulk inflow velocity. We compute the mean values over a time interval of a length corresponding to 40 cube side lengths, which we take as approximations of c_D and c_L defined as mean values over very long time.

The incoming flow is laminar time-independent with a laminar boundary layer on the front surface of the body, which separates and develops a turbulent time-dependent wake attaching to the rear of the body. The flow is thus very complex with a combination of laminar and turbulent features including boundary layers and a large turbulent wake, see Figure 17.3.

The dual problem corresponding to c_D has boundary data of unit size for φ_h on the cube in the direction of the main flow, acting on the time interval underlying the mean value, and zero boundary data elsewhere. A snapshot of the dual solution corresponding to c_D is shown in Figure 17.4, and in Figure 17.5 we plot $\|\varphi_h\|_{L_2(H^1)}$ as a function of h^{-1}, with h the smallest element diameter in the computational mesh.

We find that $\|\varphi_h\|_{L_2(H^1)}$ shows a slow logarithmic growth, and extrapolating we find that $\|\varphi_\nu\|_{L_2(H^1)} \sim \nu^{-1/2}$. We take this as evidence of computability and weak uniqueness of c_D, and we obtain similar results for the lift coefficient c_L.

[handwritten annotation in right margin: laminar & turbulent features are joined together, not separate.]

Non-Uniqueness of $D(t)$

We now investigate the computability and weak uniqueness of the total drag force $D(t)$ at a specific value time t. In Figure 17.6 we show the variation in time of $D(t)$ computed on different meshes, and we notice that $D(t)$ for a given t appears to converge very slowly or not at all with decreasing h.

We now choose one of the finer meshes corresponding to $h^{-1} \approx 500$, and we compute the dual solution corresponding to a mean value of $D(t)$ over a time interval $[T_0, T]$, where we let $T_0 \to T$. We thus seek to compute $D(T)$.

In Figure 17.7 we find a growth of $\|\varphi_h\|_{L_2(H^1)}$ similar to $|T - T_0|^{-1/2}$, as we let $T_0 \to T$. The results show that for $|T - T_0| = 1/16$ we have $\|\varphi_h\|_{L_2(H^1)} \approx 10\nu^{-1}$, and extrapolation of the computational results indicate further growth of $\|\varphi_h\|_{L_2(H^1)}$, as $T_0 \to T$ and $h \to \nu$. We take this as evidence of non-computability and weak non-uniqueness of $D(T)$.

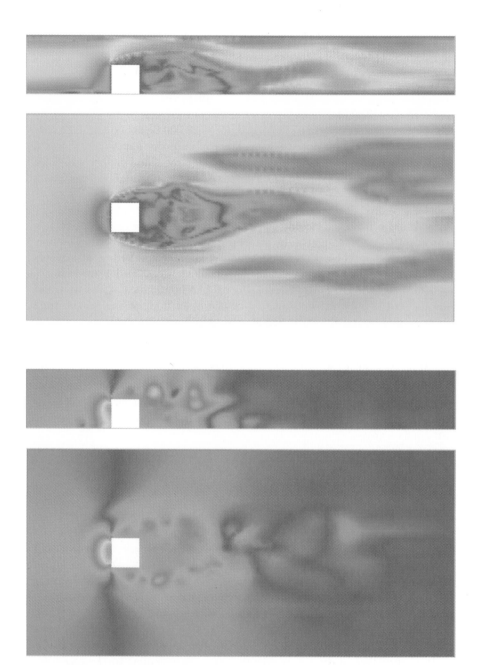

FIGURE 17.3. Velocity $|u|$ (upper) and pressure $|p|$ (lower), after 13 adaptive mesh refinements, in the x_1x_2-plane at $x_3 = 3.5H$ and in the x_1x_3-plane at $x_2 = 0.5H$, with H being the cube side length.

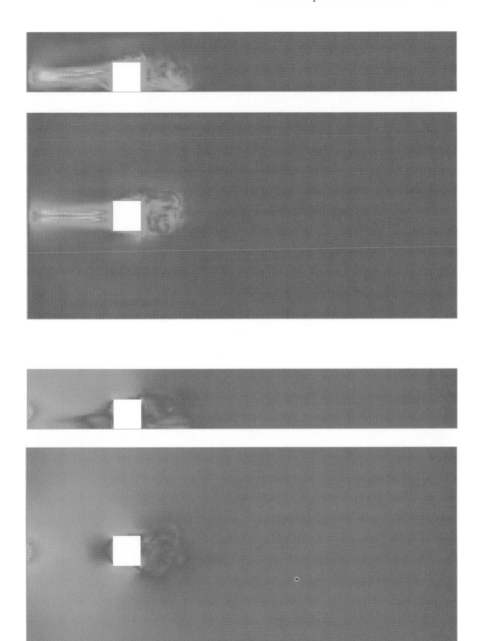

FIGURE 17.4. Dual velocity $|\varphi|$ (upper) and dual pressure $|\theta|$ (lower), after 14 adaptive mesh refinements with respect to mean drag, in the x_1x_2-plane at $x_3 = 3.5H$ and in the x_1x_3-plane at $x_2 = 0.5H$, with H being the cube side length.

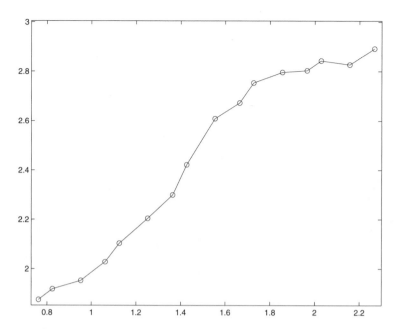

FIGURE 17.5. $\log_{10}$-$\log_{10}$-plot of $\|\varphi_h\|_{L_2(H^1)}$ as a function of $1/h$.

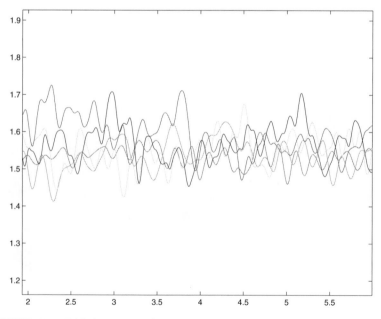

FIGURE 17.6. $D(t)$ (normalized) as a function of time, for the 5 finest computational meshes.

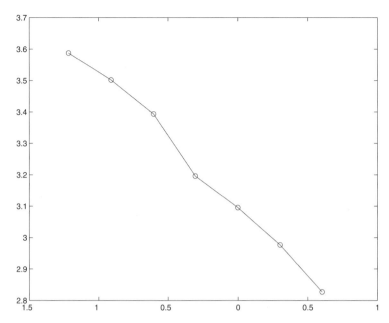

FIGURE 17.7. $\|\varphi_h\|_{L_2(H^1)}$ corresponding to computation of the mean drag force (normalized) over a time interval $[T_0, T]$, as a function of the interval length $|T - T_0|$ ($\log_{10}$-$\log_{10}$-plot).

17.14   Conclusion

We have given computational evidence of weak uniqueness of mean values such as c_D and c_L and weak non-uniqueness of a momentary value $D(t)$ of the total drag. In the computations we observe this phenomenon as a continuous degradation of computability (increasing error tolerance) as the length of the mean value decreases to zero. When the error tolerance is larger than one, then we have effectively lost computability, since the oscillation of $D(t)$ is of unit size. We compute c_D and c_L as mean values of finite length (of size 10), and thus we expect some variation also of these values, but on a smaller scale than for $D(t)$, maybe of size = 0.1 with 0.01 as a possible lower limit with present computers. Thus the distinction between computability (or weak uniqueness) and non-computability (weak non-uniqueness) may in practice be just one or two orders of magnitude in output error, rather than a difference between 0 and ∞.

Of course, this is what you may expect in a quantified computational world, as compared to an ideal mathematical world. In particular, we are led to measure residuals of approximate weak solutions, rather than working with exact weak solutions with zero residuals. A such quantified math-   Such a ...
ematical world is in fact richer than an ideal zero residual world, and thus possibly more accessible.

18
Do Mathematicians Quarrel?

The proofs of Bolzano's and Weierstrass theorems have a decidedly non-constructive character. They do not provide a method for actually finding the location of a zero or the greatest or smallest value of a function with a prescribed degree of precision in a finite number of steps. Only the mere existence, or rather the absurdity of the non-existence, of the desired value is proved. This is another important instance where the "intuitionists" have raised objections; some have even insisted that such theorems be eliminated from mathematics. The student of mathematics should take this no more seriously than did most of the critics. (Courant)

I know that the great Hilbert said "We will not be driven out from the paradise Cantor has created for us", and I reply "I see no reason to walking in". (R. Hamming)

There is a concept which corrupts and upsets all others. I refer not to the Evil, whose limited realm is that of ethics; I refer to the infinite. (Borges).

Either mathematics is too big for the human mind or the human mind is more than a machine. (Gödel)

18.1   Introduction

Mathematics is often taught as an "absolute science" where there is a clear distinction between true and false or right and wrong, which should be universally accepted by all professional mathematicians and every enlightened layman. This is true to a large extent, but there are important aspects of mathematics where agreement has been lacking and still is lacking. The development of mathematics in fact includes as fierce quarrels as any other science. In the beginning of the 20th century, the very foundations of mathematics were under intense discussion. In parallel, a split between "pure" and "applied" mathematics developed, which had never existed before. Traditionally, mathematicians were generalists combining theoretical mathematical work with applications of mathematics and even work in mechanics, physics and other disciplines. Leibniz, Lagrange, Gauss, Poincaré and von Neumann all worked with concrete problems from mechanics, physics and a variety of applications, as well as with theoretical mathematical questions.

In terms of the foundations of mathematics, there are different "mathematical schools" that view the basic concepts and axioms somewhat differently and that use somewhat different types of arguments in their work. The three principal schools are the *formalists*, the *logicists* and finally the *intuitionists*, also known as the *constructivists*.

As we explain below, we group both the formalists and the logicists together under an *idealistic* tradition and the the the constructivists under a *realistic* tradition. It is possible to associate the idealistic tradition to an "aristocratic" standpoint and the realistic tradition to a "democratic" one. The history of the Western World can largely be be viewed as a battle between an idealistic/aristocratic and a realistic/democratic tradition. The Greek philosopher Plato is the portal figure of the idealistic/aristocratic tradition, while along with the scientific revolution initiated in the 16th century, the realistic/democratic tradition has taken a leading role in our society.

The debate between the formalists/logicists and the constructivists culminated in the 1930s, when the program put forward by the formalists and logicists suffered a strong blow from the logician Kurt Gödel. Gödel showed, to the surprise of world including great mathematicians like Hilbert, that in any axiomatic mathematical theory containing the axioms for the natural numbers, there are true facts which cannot be proved from the axioms. This is Gödel's famous *incompleteness theorem*.

Alan Turing (1912-54, dissertation at Kings College, Cambridge 1935) took up a similar line of thought in the form of computability of real numbers in his famous 1936 article *On Computable Numbers, with an application to the Entscheidungsproblem*. In this paper Turing introduced an abstract machine, now called a *Turing machine*, which became the prototype of the modern programmable computer. Turing defined a computable number as real number whose decimal expansion could be produced by a Turing machine. He showed that π was computable, but claimed that most real numbers are not computable. He gave gave examples of "undecidable problems" formulated as the problem if the Turing machine would come to a halt or not, see Fig. 18.2. Turing laid out plans for an electronic computer named Analytical Computing Engine ACE, with reference to Babbage's' Analytical Engine, at the same time as the ENIAC was designed in the US.

FIGURE 18.1. Kurt Gödel (with Einstein 1950): "Every formal system is incomplete."

Gödel's and Turing's work signified a clear defeat for the formalists/ logicists and a corresponding victory for the constructivists. Paradoxically, soon after the defeat the formalists/logicists gained control of the mathematics departments and the constructivists left to create new departments of computer science and numerical analysis based on constructive mathematics. It appears that the trauma generated by Gödel's and Turing's findings on the incompleteness of axiomatic methods and un-computability, was so strong that the earlier co-existence of the formalists/logicists and constructivists was no longer possible. Even today, the world of mathematics is heavily influenced by this split.

We will come back to the dispute between the formalists/logicists and constructivists below, and use it to illustrate fundamental aspects of mathematics which hopefully can help us to understand our subject better.

1937: Alan Turing's theory of digital
computing

FIGURE 18.2. Alan Turing: "I wonder if my machine will come to a halt.".

18.2   The Formalists

The *formalist* school says that it does not matter what the basic concepts
actually mean, because in mathematics we are just concerned with relations
between the basic concepts whatever the meaning may be. Thus, we do
not have to (and cannot) explain or define the basic concepts and can view
mathematics as some kind of "game". However, a formalist would be very
anxious to demonstrate that in his formal system it would not be possible
to arrive at *contradictions*, in which case his game would be at risk of
breaking down. A formalist would thus like to be absolutely sure about the
consistency of his formal system. Further, a formalist would like to know
that, at least in principle, he would be able to understand his own game
fully, that is that he would in principle be able to give a mathematical
explanation or proof of any true property of his game. The mathematician
Hilbert was the leader of the formalist school. Hilbert was shocked by the
results by Gödel. Did he ever accept them?

18.3   The Logicists and Set Theory

The logicists try to base mathematics on logic and *set theory*. Set theory
was developed during the second half of the 19th century and the language
of set theory has become a part of our every day language and is very
much appreciated by both the formalist and logicist schools, while the

FIGURE 18.3. Bertrand Russell: "I am protesting."

constructivists have a more reserved attitude. A set is a collection of items, which are the elements of the set. An element of the set is said to belong to the set. For example, a dinner may be viewed as a set consisting of various dishes (entree, main course, dessert, coffee). A family (the Wilsons) may be viewed as a set consisting of a father (Mr. Wilson), a mother (Mrs. Wilson) and two kids (Tom and Mary). A soccer team (IFK Göteborg for example) consists of the set of players of the team. Humanity may be said to be set of all human beings.

Set theory makes it possible to speak about collections of objects as if they were single objects. This is very attractive in both science and politics, since it gives the possibility of forming new concepts and groups in hierarchical structures. Out of old sets, one may form new sets whose elements are the old sets. Mathematicians like to speak about *the set of all real numbers*, denoted by $\mathbb{R}$, *the set of all positive real numbers*, and the *set of all prime numbers*, and a politician planning a campaign may think of the set of democratic voters, the set of auto workers, the set of female first time voters, or the set of all poor, jobless, male criminals. Further, a workers union may be thought of as a set of workers in a particular factory or field, and workers unions may come together into unions or sets of workers unions.

A set may be described by listing all the elements of the set. This may be very demanding if the set contains many elements (for example if the set is humanity). An alternative is to describe the set through a property shared by all the elements of the set, e.g. the set of all people who have the properties of being poor, jobless, male, and criminal at the same time. To describe humanity as the set of beings which share the property of being

yes { human, however seems to more of a play with words than something very
useful.

The leader of the logicist school was the philosopher and peace activist
Bertrand Russell (1872-1970). Russell discovered that building sets freely
can lead into contradictions that threaten the credibility of the whole logicist system. Russell created variants of the old *liar's paradox* and *barber's
paradox*, which we now recall. Gödel's theorem may be viewed to a variant
of this paradox.

The Liar's Paradox

The liar's paradox goes as follows: A person says "I am lying". How should
you interpret this sentence? If you assume that what the person says is
indeed true, then it means that he is lying and then what he says is not
true. On the other hand, if you assume that what he says is not true, this
means that he is not lying and thus telling the truth, which means that
what he says is true. In either case, you seem to be led to a contradiction,
right? Compare Fig. 18.4.

FIGURE 18.4. "I am (not) lying"

The Barber's Paradox

The barber's paradox goes as follows: The barber in the village has decided
to cut the hair of everyone in the village who does not cut his own hair.
What shall the barber do himself? If he decides to cut his own hair, he will
belong to the group of people who cut their own hair and then according to
his decision, he should not cut his own hair, which leads to a contradiction.

On the other hand, if he decides not to cut his own hair, then he would belong to the group of people not cutting their own hair and then according to his decision, he should cut his hair, which is again a contradiction. Compare Fig. 18.5.

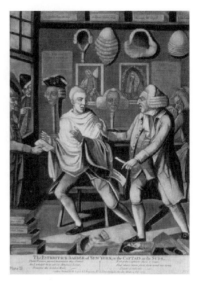

FIGURE 18.5. Attitudes to the "barber's paradox": one relaxed and one very concerned.

18.4   The Constructivists

The *intuitionist/constructivist* view is to consider the basic concepts to have a meaning which may be directly "intuitively" understood by our brains and bodies through experience, without any further explanation. Furthermore, the intuitionists would like to use as concrete or "constructive" arguments as possible, in order for their mathematics always to have an intuitive "real" meaning and not just be a formality like a game.

Basically I agree with this.

An intuitionist may say that the natural numbers 1, 2, 3,...., are obtained by repeatedly adding 1 starting at 1. We know that from the constructivist point of view, the natural numbers are something in the state of being created in a process without end. Given a natural number n, there is always a next natural number $n + 1$ and the process never stops. A constructivist would not speak of the set of all natural numbers as something having been completed and constituting an entity in itself, like the set of all natural numbers as a formalist or logicist would be willing to do. Gauss pointed out that "the set of natural numbers" rather would reflect a "mode of speaking" than existence as a set.

FIGURE 18.6. Luitzen Egbertus Jan Brouwer 1881-1966: :"One cannot inquire into the foundations and nature of mathematics without delving into the question of the operations by which mathematical activity of the mind is conducted. If one failed to take that into account, then one would be left studying only the language in which mathematics is represented rather than the essence of mathematics".

An intuitionist would not feel a need of "justification" or a proof of consistency through some extra arguments, but would say that the justification is built into the very process of developing mathematics using constructive processes. A constructivist would so to speak build a machine that could fly (an airplane) and that very constructive process would itself be a proof of the claim that building an airplane would be possible. A constructivist is *Yes* thus in spirit close to a practicing engineer. A formalist would not actually build an airplane, rather make some model of an airplane, and would then need some type of argument to convince investors and passengers that his airplane would actually be able to fly, at least in principle. The leader of the intuitionist school was Brouwer (1881-1966), see Fig. 18.6. Hard-core constructivism makes life very difficult (like strong vegetarianism), and because the Brouwer school of constructivists were rather fundamentalist in their spirit, they were quickly marginalized and lost influence in the 1930s. The quote by Courant given above shows the strong feelings involved related to the fact that very fundamental dogmas were at stake, and the general lack of rational arguments to meet the criticism from the intuitionists, which was often replaced by ridicule and oppression.

Van der Waerden, a mathematician who studied at Amsterdam from 1919 to 1923 wrote: "Brouwer came [to the university] to give his courses but lived in Laren. He came only once a week. In general that would have not been permitted - he should have lived in Amsterdam - but for him an exception was made. ... I once interrupted him during a lecture to ask a question. Before the next week's lesson, his assistant came to me to say that Brouwer did not want questions put to him in class. He just did not want them, he was always looking at the blackboard, never towards the students. ... Even though his most important research contributions were in topology, Brouwer never gave courses on topology, but always on - and only on - the foundations of intuitionism . It seemed that he was no longer convinced of his results in topology because they were not correct from the point of view of intuitionism, and he judged everything he had done before, his greatest output, false according to his philosophy. He was a very strange person, crazy in love with his philosophy".

18.5 The Peano Axiom System for Natural Numbers

The Italian mathematician Peano (1858-1932) set up an axiom system for the natural numbers using as undefined concepts "natural number", "successor", "belong to", "set" and "equal to". His five axioms are

1. 1 is a natural number

2. 1 is not the successor of any other natural number   *What about 1 being successor to ∅?*

3. Each natural number n has a successor *(More than one or only one successor?)*

4. If the successors of n and m are equal then so are n and m

There is a fifth axiom which is the axiom of *mathematical induction* stating that if a property holds for any natural number n, whenever it holds for the natural number preceding n and it holds for $n = 1$, then it holds for *all natural numbers.* Starting with these five axioms, one can derive all the basic properties of real numbers.

We see that the Peano axiom system tries to catch the essence of our intuitive feeling of natural numbers as resulting from successively adding 1 without ever stopping. The question is if we get a more clear idea of the natural numbers from the Peano axiom system than from our intuitive feeling. Maybe the Peano axiom system helps to identify the basic properties of natural numbers, but it is not so clear what the improved insight really consists of. *Amen!*

The logicist Russell proposed in *Principia Mathematica* to define the natural numbers using set theory and logic. For instance, the number 1 would be defined roughly speaking as the set of all singletons, the number two the set of all dyads or pairs, the number three as the set of all triples, et cet. Again the question is if this adds insight to our conception of natural numbers? *With this definition "one" becomes a class, a fairly complicated and logically elaborate class.*

18.6   Real Numbers

Many textbooks in calculus start with the assumption that the reader is already familiar with *real numbers* and quickly introduce the notation $\mathbb{R}$ to denote the set of *all real numbers*. The reader is usually reminded that the real numbers may be represented as points on the *real line* depicted as a horizontal (thin straight black) line with marks indicating 1, 2, and maybe numbers like 1.1, 1.2, $\sqrt{2}$, and π. This idea of basing *arithmetic*, that is numbers, on *geometry* goes back to Euclid, who took this route to get around the difficulties of irrational numbers discovered by the Pythagoreans. However, relying solely on arguments from geometry is very impractical and Descartes turned the picture around in the 17th century by basing geometry on arithmetic, which opened the way to the revolution of Calculus. The difficulties related to the evasive nature of irrational numbers encountered by the Pythagoreans, then of course reappeared, and the related questions concerning the very foundations of mathematics gradually developed into a quarrel with fierce participation of many of the greatest mathematicians which culminated in the 1930s, and which has shaped the mathematical world of today.

We have come to the standpoint above that a real number may be defined through its decimal expansion. A rational real number has a decimal expansion that eventually becomes periodic. An irrational real number has an expansion which is infinite and is not periodic. We have defined $\mathbb{R}$ as the set of all possible infinite decimal expansions, with the agreement that this definition is a bit vague because the meaning of "possible" is vague. We may say that we use a constructivist/intuitionist definition of $\mathbb{R}$.

The formalist/logicist would rather like to define $\mathbb{R}$ as the set of all infinite decimal expansions, or set of all Cauchy sequences of rational numbers, in what we called a universal Big Brother style above.

The set of real numbers is often referred to as the "continuum" of real numbers. The idea of a "continuum" is basic in classical mechanics where both space and time is supposed to be "continuous" rather than "discrete". On the other hand, in quantum mechanics, which is the modern version of mechanics on the scales of atoms and molecules, matter starts to show features of being discrete rather than continuous. This reflects the famous particle-wave duality in quantum mechanics with the particle being discrete and the wave being continuous. Depending on what glasses we use, phenomena may appear to be more or less discrete or continuous and no single mode of description seems to suffice. The discussions on the nature of real numbers may be rooted in this dilemma, which may never be resolved.

[handwritten left margin: This is the first time they use the phrase "Big Brother style". They don't use it above.]

[handwritten right margin: A very arguable proposition.]

[handwritten bottom: → It is rooted in this dilemma. This is the paradox of Achilles and the tortoise, again.]

18.7   Cantor versus Kronecker

Let us give a glimpse of the discussion on the nature of real numbers through two of the key personalities, namely Cantor (1845-1918) in the *formalist* corner and Kronecker (1823-91), in the *constructivist* corner. These two mathematicians were during the late half of the 19th century involved in a bitter academic fight through their professional lives (which eventually led Cantor into a tragic mental disorder). Cantor created *set theory* and in particular a theory about sets with *infinitely* many elements, such as the set of natural numbers or the set of real numbers. Cantors theory was criticized by Kronecker, and many others, who simply could not believe in Cantors mental constructions or consider them to be really interesting. Kronecker took a down-to-earth approach and said that only sets with finitely many elements can be properly understood by human brains ("God created the integers, all else is the work of man"). Alternatively, Kronecker said that only mathematical objects that can be "constructed" in a *finite* number of steps actually "exist", while Cantor allowed infinitely many steps in a "construction". Cantor would say that the set of *all natural numbers* that is the set with the elements $1, 2, 3, 4, ...$, would "exist" as an object in itself as *the set of all natural numbers* which could be grasped by human brains, while Kronecker would deny such a possibility and reserve it to a higher being. Of course, Kronecker did not claim that there are only finitely many natural numbers or that there is a largest natural number, but he would (following Aristotle) say that the existence of arbitrarily large natural numbers is like a "potential" rather than an actual reality.

FIGURE 18.7. Cantor (left): "I realize that in this undertaking I place myself in a certain opposition to views widely held concerning the mathematical infinite and to opinions frequently defended on the nature of numbers". Kronecker (right): "God created the integers, all else is the work of man".

In the first round, Kronecker won since Cantor's theories about the infinite was rejected by many mathematicians in the late 19th and beginning 20th century. But in the next round, the influential mathematician Hilbert, the leader of the formalist school, joined on the side of Cantor. Bertrand Russell and Norbert Whitehead tried to give mathematics a foundation based on logic and set theory in their monumental *Principia Mathematica* (1910-13) and may also be viewed as supporters of Cantor. Thus, despite the strong blow from Gödel in the 1930's, the formalist/logicist schools took over the scene and have been dominating mathematics education into our time. Today, the development of the computer is again starting to shift the weight to the side of the constructivists, simply because no computer is able to perform infinitely many operations nor store infinitely many numbers, and so the old battle may come alive again.

Cantor's theories about infinite numbers have mostly been forgotten, but there is one reminiscence in most presentations of the basics of Calculus, namely Cantor's argument that the degree of infinity of the real numbers is strictly larger than that of the rational or natural numbers. Cantor argued as follows: suppose we try to enumerate the real numbers in a list with a first real number r_1, a second real number r_2 and so on. Cantor claimed that in any such list there must be some real numbers missing, for example any real number that differs from r_1 in the first decimal, from r_2 in the second decimal and so on. Right? Kronecker would argue against this construction simply by asking full information about for example r_1, that is, full information about all the digits of r_1. OK, if r_1 was rational then this could be given, but if r_1 was irrational, then the mere listing of all the decimals of r_1 would never come to an end, and so the idea of a list of real numbers would not be very convincing. So what do you think? Cantor or Kronecker?

Good "point"

Cantor not only speculated about different degrees of infinities, but also cleared out more concrete questions about e.g. convergence of trigonometric series viewing real numbers as limits of of Cauchy sequences of rational numbers in pretty much the same we have presented.

18.8   Deciding Whether a Number is Rational or Irrational

We dwell a bit more on the nature of real numbers. Suppose x is a real number, the decimals of which can be determined one by one by using a certain algorithm. How can we tell if x is rational or irrational? Theoretically, if the decimal expansion is periodic then x is rational otherwise it is irrational. There is a practical problem with this answer however because we can only compute a finite number of digits, say never more than 10^{100}. How can we be sure that the decimal expansion does not start repeating af-

Suppose the period is very long: Suppose the period is very long?

24 Oct 2005

⌐ This relates to Turing's halting problem. When does⌐

ter that? To be honest, this question seems very difficult to answer. Indeed
it appears to be impossible to tell what happens in the complete decimal
expansion by looking at a finite number of decimals. The only way to decide if a number x is rational or irrational is figure out a clever argument
like the one the Pythagoreans used to show that $\sqrt{2}$ is irrational. Figuring
out such arguments for different specific numbers like π and e is an activity
that has interested a lot of mathematicians over the years.

On the other hand, the computer can only compute rational numbers and
moreover only rational numbers with finite decimal expansions. If irrational
numbers do not exist in practical computations, it is reasonable to wonder if
they truly exist. Constructive mathematicians like Kronecker and Brouwer
would not claim that irrational numbers really exist. *I tend to agree.*

18.9   The Set of All Possible Books

We suggest it is reasonable to define the set of all real numbers $\mathbb{R}$ as *the
set of all possible decimal expansions* or equivalently *the set of all possible Cauchy sequences of rational numbers*. Periodic decimal expansions
correspond to rational numbers and non-periodic expansions to irrational
numbers. The set $\mathbb{R}$ thus consists of the set of all rational numbers together
with the set of all irrational numbers. We know that it is common to omit
the word "possible" in the suggested definition of $\mathbb{R}$ and define $\mathbb{R}$ as "the
set of all real numbers", or "the set of all infinite decimal expansions".

Let's see if this hides some tricky point by way of an analogy. Suppose
we define a "book" to be any finite sequence of letters. There are specific
books such as "The Old Man and the Sea" by Hemingway, "The Author
as a Young Dog" by Thomas, "Alice in Wonderland" by Lewis Carrol, and
"1984" by Orwell, that we could talk about. We could then introduce **B** as
"the set of all possible books", which would consist of all the books that
have been and will be written purposely, together with many more "books"
that consist of random sequences of letters. These would include those famous books that are written or could be written by chimpanzees playing
with typewriters. We could probably handle this kind of terminology without too much difficulty, and we would agree that 1984 is an element of **B**.
More generally, we would be able to say that any given book is a member
of **B**. Although this statement is difficult to deny, it is also hard to say that
this ability is very useful.

Suppose now we omit the word possible and start to speak of **B** as "the
set of all books". This could give the impression that in some sense **B** is an
existing reality, rather than some kind of potential as when we speak about
"possible books". The set **B** could then be viewed as a library containing
all books. This library would have to be enormously large and most of the

This, of course, is Borges' library

the first period halt x & the second period begin? Suppose there is no third or fourth period.

"books" would be of no interest to anyone. Believing that the set of all books "exists" as a reality would not be very natural for most people.

The set of real numbers $\mathbb{R}$ has the same flavor as the set of all books **B**. It must be a very large set of numbers of which only a relative few, such as the rational numbers and a few specific irrational numbers, are ever encountered in practice. Yet, it is traditional to define $\mathbb{R}$ as the set of real numbers, rather than as "set of all possible real numbers". The reader may choose the interpretation of $\mathbb{R}$ according to his own taste. A true idealist would claim that the set of all real numbers "exists", while a down-to-earth person would more likely speak about the set of possible real numbers. Eventually, this may come down to a personal religious feeling; some people appear to believe that Heaven actually exists, and while others might view as a potential or as a poetic way of describing something which is difficult to grasp.

Whatever interpretation you choose, you will certainly agree that some real numbers are more clearly specified than others, and that to specify a real number, you need to give some algorithm allowing you to determine as many digits of the real number as would be possible (or reasonable) to ask for.

18.10   Recipes and Good Food

Using the Bisection algorithm, we can compute any number of decimals of $\sqrt{2}$ if we have enough computational power. Using an algorithm to specify a number is analogous to using a recipe to specify for example *Grandpa's Chocolate Cake*. By following the recipe, we can bake a cake that is a more or less accurate approximation of the ideal cake (which only Grandpa can make) depending on our skill, energy, equipment and ingredients. There is a clear difference between the recipe and cakes made from the recipe, since after all we can eat a cake with pleasure but not a recipe. The recipe is like an algorithm or scheme telling us how to proceed, how many eggs to use for example, while cakes are the result of actually applying the algorithm with real eggs.

Of course, there are people who seem to enjoy reading recipes, or even just looking at pictures of food in magazines and talking about it. But if they never actually do cook anything, their friends are likely to lose interest in this activity. Similarly, you may enjoy looking at the symbols π or $\sqrt{2}$ and talking about them, or writing them on pieces of paper, but if you never actually compute them, you may come to wonder what you are actually doing.

In this book, we will see that there are many mathematical quantities that can only be determined approximately using a computational algorithm. Examples of such quantities are $\sqrt{2}$, π, and the base e of the natural

[handwritten marginalia, left:] I much prefer the use of the word "possible". HA!

[handwritten marginalia, right:] Just as some animals are more equal than others.

logarithm. Later we will find that there are also functions, even elementary functions like $\sin(x)$ and $\exp(x)$ that need to be computed for different values of x. Just as we first need to bake a cake in order to enjoy it, we may need to compute such ideal mathematical quantities using certain algorithms before using them for other purposes.

18.11   The "New Math" in Elementary Education

After the defeat of formalists in the 1930s by the arguments of Gödel, paradoxically the formalist school took over and set theory got a new chance. A wave generated by this development struck the elementary mathematics education in the 1960s in the form of the "new math". The idea was to explain numbers using set theory, just as Russell and Whitehead had tried to do 60 years earlier in their *Principia*. Thus a kid would learn that a set consisting of one cow, two cups, a piece of chocolate and an orange, would have five elements. The idea was to explain the nature of the number 5 this way rather than counting to five on the fingers or pick out 5 oranges from a heap of oranges. This type of "new math" confused the kids, and the parents and teachers even more, and was abandoned after some years of turbulence. And maybe real numbers are the source of turbulence.

One might say a tsunami wave, and probably failed.

18.12   The Search for Rigor in Mathematics

The formalists tried to give mathematics a rigorous basis. The search for rigor was started by Cauchy and Weierstrass who tried to give precise definitions of the concepts of limit, derivative and integral, and was continued by Cantor and Dedekind who tried to clarify the precise meaning of concepts such as continuum, real number, the set of real numbers, and so on. Eventually this effort of giving mathematics a fully rational basis collapsed, as we have indicated above.

We may identify two types of rigor: *May we identify two types of collapse, logical & practical?*

- constructive rigor

- formal rigor.

Constructive rigor is necessary to accomplish difficult tasks like carrying out a heart operation, sending a man to the moon, building a tall suspension bridge, climbing Mount Everest, or writing a long computer program that works properly. In each case, every little detail may count and if the whole enterprise is not characterized by extreme rigor, it will most likely fail. *Yes* Eventually this is a rigor that concerns material things, or real events.

Formal rigor is of a different nature and does not have a direct concrete objective like the ones suggested above. Formal rigor may be exercised at a royal court or in diplomacy, for example. It is a rigor that concerns language (words), or manners. The Scholastic philosophers during the Medieval time, were formalists who loved formal rigor and could discuss through very complicated arguments for example the question how many Angels could fit onto the edge of a knife. Some people use a very educated formally correct language which may be viewed as expressing a formal rigor. Authors pay a lot of attention to the formalities of language, and may spend hour after hour polishing on just one sentence until it gets just the right form. More generally, formal aspects may be very important in Arts and Aesthetics. Formal rigor may be thus very important, but serves a different purpose than constructive rigor. Constructive rigor is there to guarantee that something will actually function as desired. Formal rigor may serve the purpose of controlling people or impressing people, or just make people feel good, or to carry out a diplomatic negotiation. Formal rigor may be exercised in a game or play with certain very specific rules, that may be very strict, but do not serve a direct practical purpose outside the game.

Also in mathematics, one may distinguish between concrete and formal error. A computation, like multiplication of two natural numbers, is a concrete task and rigor simply means that the computation is carried out in a correct way. This may be very important in economics or engineering. It is not difficult to explain the usefulness of this type of constructive rigor, and the student has no difficulty in formulating himself what the criteria of constructive rigor might be in different contexts.

Formal rigor in calculus was promoted by Weierstrass with the objective of making basic concepts and arguments like the continuum of real numbers or limit processes more "formally correct". The idea of formal rigor is still alive very much in mathematics education dominated by the formalist school. Usually, students cannot understand the meaning of this type of "formally rigorous reasoning", and very seldom can exercise this type of rigor without much direction from the teacher.

We shall follow an approach where we try to reach constructive rigor to a degree which can be clearly motivated, and we shall seek to make the concept of formal rigor somewhat understandable and explain some of its virtues.

Formal vs. constructive rigor: [handwritten marginal note]

18.13   A Non-Constructive Proof

We now give an example of a proof with non-constructive aspects that plays an important role in many Calculus books. Although because of the non-constructive aspects, the proof is considered to be so difficult that it can only by appreciated by selected math majors. *Yeah, right!* [handwritten note]

The setting is the following: We consider a bounded increasing sequence $\{a_n\}_1^\infty$ of real numbers, that is $a_n \leq a_{n+1}$ for $n = 1, 2, ...$, and there is a constant C such that $a_n \leq C$ for $n = 1, 2, ...$ The claim is that the sequence $\{a_n\}_1^\infty$ converges to a limit A. The proof goes as follows: all the numbers a_n clearly belong to the interval $I = [a_1, C]$. For simplicity suppose $a_1 = 0$ and $C = 1$. Divide now the interval $[0, 1]$ into the two intervals $[0, 1/2]$ and $[1/2, 1]$. and make the following choice: if there is a real number a_n such that $a_n \in [1/2, 1]$, then choose the right interval $[1/2, 1]$ and if not choose the left interval $[0, 1/2]$. Then repeat the subdivision into a left and a right interval, choose one of the intervals following the same principle: if there is a real number a_n in the right interval, then choose this interval, and if not choose the left interval. We then get a nested sequence of intervals with length tending to zero defining a unique real number that is easily seen to be the limit of the sequence $\{a_n\}_1^\infty$. Are you convinced? If not, you must be a constructivist.   *Yes, I must be a constructivist.*

So where is the hook of non-constructiveness in this proof? Of course, it concerns the choice of interval: in order to choose the correct interval you must be able to check if there is some a_n that belongs to the right interval, that is you must check if a_n belongs to the right interval for all sufficiently large n. The question from a constructivist point of view is if we can perform each check in a finite number of steps. Well, this may depend on the particular sequence $a_n \leq a_{n+1}$ under consideration. Let's first consider a sequence which is so simple that we may say that we know everything of interest: for example the sequence $\{a_n\}_1^\infty$ with $a_n = 1 - 2^{-n}$, that is the sequence $1/2, 3/4, 7/8, 15/16, 31/32, ...$, which is a bounded increasing sequence clearly converging to 1. For this sequence, we would be able to always choose the correct interval (the right one) because of its simplicity.

We now consider the sequence $\{a_n\}_1^\infty$ with $a_n = \sum_1^n \frac{1}{k^2}$, which is clearly an increasing sequence, and one can also quite easily show that the sequence is bounded. In this case the choice of interval is much more tricky, and it is not clear how to make the choice constructively without actually constructing the limit. So there we stand, and we may question the value of the non-constructive proof of existence of a limit, if we anyway have to construct the limit. In any case we sum up in the following result:

Theorem 18.1 (non-constructive!) *A bounded increasing sequence converges.*

18.14   Summary

The viewpoint of Plato was to say that ideal points and lines exist in some Heaven above, while the points and lines which we as human beings can deal with, are some more or less incomplete copies or shades or images of the ideals. This is Plato's *idealistic* approach, which is related to the formalistic school. An intuitionist would say that we can never be sure of the existence of the ideals, and that we should concentrate on the more

[margin handwriting: Not totally convinced because there is a potential infinite vagueness in here. I.e. "with length tending to zero". We can't prove the interval lengths tending to zero.]

[bottom handwriting: Just assume H. tends to zero. We have to see it. tending to zero.]

or less incomplete copies we can *construct* ourselves as human beings. The question of the actual existence of the ideals thus becomes a question of *metaphysics* or *religion*, to which there probably is no definite answer. Following our own feelings, we may choose to be either a idealist/formalist or an intuitionist/constructivist, or something in between.

Yes, we may choose

The authors of this book have chosen such a middle way between the constructivist and formalist schools, trying always to be as constructive as is possible from a practical point of view, but often using a formalist language for reasons of convenience. The constructive approach puts emphasis on the concrete aspects of mathematics and brings it close to engineering and "body". This reduces the mystical character of mathematics and helps understanding. On the other hand, mathematics is not equal to engineering or only "body", and also the less concrete aspects or "soul" are useful for our thinking and in modeling the world around us. We thus seek a good synthesis of constructive and formalistic mathematics, or a synthesis of Body & Soul.

Going back to the start of our little discussion, we thus associate the logicist and formalistic schools with the idealistic/aristocratic tradition and the constructivists with the constructive/democratic tradition. As students, we would probably appreciate a constructive/democratic approach, since it aids the understanding and gives the student an active role. On the other hand, certain things indeed are very difficult to understand or construct, and then the idealistic/aristocratic approach opens a possible attitude to handle this dilemma.

The constructivist approach, whenever feasible, is appealing from educational point of view, since it gives the student an active role. The student is invited to construct himself, and not just watch an omnipotent teacher pick ready-made examples from Heaven.

Summary of their position

Of course, the development of the modern computer has meant a tremendous boost of constructive mathematics, because what the computer does is constructive. Mathematics education is still dominated by the formalist school, and the most of the problems today afflicting mathematics education can be related to the over-emphasis of the idealistic school in times when constructive mathematics is dominating in applications.

Turing's principle of a "universal computing machine" directly connects the work on the foundations of mathematics in the 1930s (with *Computable numbers* as a key article), with the development of the modern computer in the 1940s (with ACE as a key example), and thus very concretely illustrates the power of (constructive!) mathematics.

Some distinguished mathematicians have recently advocated the more or less complete banishment from mathematics of all non-constructive proofs. Even if such a program were desirable, it would involve tremendous complications and even the partial destruction of the body of living mathematics. For this reason it is no wonder that the school of "intuitionism", which has adopted this program, has met with strong resistance, and that even the most thoroughgoing intuitionists cannot always live up to their convictions. (Courant)

The composition of vast books is a laborious and impoverishing extravagance. To go on for five hundred pages developing an idea whose perfect oral exposition is possible in a few minutes! A better course of procedure is to pretend that these books already exist, and then to offer a resume, a commentary...More reasonable, more inept, more indolent, I have preferred to write notes upon imaginary books. (Borges, 1941)

I have always imagined that Paradise will be kind of a library. (Borges)

My prize book at Sherbourne School (von Neumann's Mathematische Grundlagen der Quantenmechanik) is turning out very interesting, and not at all difficult reading, although the applied mathematicians seem to find it rather strong. (Turing, age 21) *Boy! He must have been really good at maths as the Brits would say.*

FIGURE 18.8. View of the river Cam at Cambridge 2003 with ACE in the fore-ground (and "UNTHINKABLE" in the background to the right)

24 Oct 2005

Part III

Appendix

Appendix A
Comments on the Directives of the Mathematics Delegation

We now return to the Mathematics Delegation and shortly comment on its directives.

A.1 The Directives

- The importance of good knowledge in mathematics is undeniable. This covers a wide area from daily knowledge to creating the conditions for life-long learning, as well as the acquisition of competence and problem solving skills required for learning in other subjects, and for actively participating in society and working life.

- General skills such as logical thinking, the ability to abstract, analyse arguments, communication and problem solving skills are all developed, applied and trained within mathematics.

- For this reason, it is natural that mathematics is one of the basic subjects in the compulsory school, that the admissions requirement to the upper secondary school is a Pass Grade, and that mathematics is a core subject in the upper secondary school.

- A knowledge of mathematics helps us to understand complex contexts, and is a prerequisite not only for our joint welfare, but also the individual's opportunities to e.g. be able to examine and evaluate arguments in the political debate on the use and distribution of our joint resources.

- People with good knowledge in the natural sciences and technology are of vital importance for Sweden to be able to continue to develop as a leading industrial nation using resources effectively and to promote sustainable economic, social and ecological development. A good knowledge of mathematics is also needed in many other areas in order to achieve success. Mathematics and its applications contribute to development in a large number of areas such as e.g. electronics, communication, economics, biology and medicine, as well as the arts, music and film.

A.2   Comments

We now examine the above statements concerning form, content, and logic, possibly using some of our analytical skills developed during our studies of mathematics (and languages and other subjects).

General and Vague

The statements are general in nature, and no indication to *what* mathematics is referred to: Mathematics in general? Calculus? Linear algebra? Elementary arithmetic? Algebraic topology? And no indication at all is given to the many important applications of computational mathematics in our information society.

Problem Solving Skills

It is claimed that skills of "problem solving" are developed within mathematics education. Again within which parts of mathematics education: General? Calculus? Any? And what types of problems?

Logical Thinking and the Ability to Abstract

It is claimed that "general skills such as logical thinking, the ability to abstract and analyse arguments, are all developed, applied and trained within mathematics". It is not mentioned that such skills equally much, or even more, are developed within studies in languages (in particular the mother tongue) and in social and natural sciences.

The Undeniable Importance of Mathematics

It is stated that "The importance of good knowledge in mathematics is undeniable". Again the statement is very vague: what is "good knowledge" and what mathematics is referred to? And of course, the use of the phrase "undeniable" in an argument may be questioned. The only statements which are undeniable are tautologies like "there are 100 centimeters on a meter" or "one plus one is two".

Maybe a "good knowledge" of addition, subtraction and multiplication, is sufficient? If this is true, how could we motivate students to go beyond elementary arithmetics?

The Undeniable Importance of Latin

Similarly, one may argue from an "undeniable importance" of language that classical Latin and Greek should be reintroduced as a compulsory subject from basic education and up. Or from the "undeniable success" of pop music that the education in contemporary electronic music (appreciated by only a very small part of the public), should be significantly strengthened.

Nothing About Computational Mathematics

A most remarkable feature of the directives is that nothing is mentioned about the role and importance of computational mathematics today, in science, medicine, engineering, economics and other areas, of which we present evidence in this book. In particular no indication is given of the importance of computational mathematics within studies in natural sciences and technology, viewed to be of "vital importance for Sweden to be able to continue to develop as a leading industrial nation using resources effectively and to promote sustainable economic, social and ecological development". This is truly very remarkable!

Distribution of Joint Resources

It is claimed that mathematics helps us to evaluate arguments for the distribution of joint resources. Maybe so, but this issue in general has more to do with political ideologies than mere counting.

How to be Successful?

It is stated that "A good knowledge of mathematics is also needed in many other areas in order to achieve success such as communication, economics, biology and medicine, as well as the arts, music and film". Is this true? Maybe in economics, but is it true in the arts, music and film? Which mathematics is here useful to reach success? That $1 million plus $1 million is $2 millions?

A.3   A Sum Up

The directives are surprisingly vague and sweeping and in fact appear to reflect quite a bit of confusion (or is it a careful calculation?). It is easy to get the impression that the directives are formulated so as to:

Maybe it is careful obfuscation.

- be politically correct ("sustainable economic, social and ecological development");

- suit traditional schools of mathematics by suggesting usefulness of mathematics in general, but not giving any concretion concerning in particular the role of computational mathematics today;

- suit traditional schools of mathematics didactics relieving them from contacts with modern computational mathematics.

Of course, with such directives there is a high risk that also the report of the delegation will come out as being vague and confused. Further, there is a risk that such a report will be used to give a justification of status quo, rather than as a tool for constructive change to the better.

We invite the reader to analyze the directives, and in particular present possible reasons for the absence of computational mathematics. And of course to read and analyze the report on www.matematikdelegationen.gov.se when available.

24 Oct 2005

They mean concrete examples, but the use (or misuse) of the word concretion is rather nice.

Appendix B
Preface to Body&Soul

The Need of Reform of Mathematics Education

Mathematics education needs to be reformed as we now pass into the new millennium. We share this conviction with a rapidly increasing number of researchers and teachers of both mathematics and topics of science and engineering based on mathematical modeling. The reason is of course the computer revolution, which has fundamentally changed the possibilities of using mathematical and computational techniques for modeling, simulation and control of real phenomena. New products and systems may be developed and tested through computer simulation on time scales and at costs which are orders of magnitude smaller than those using traditional techniques based on extensive laboratory testing, hand calculations and trial and error.

At the heart of the new simulation techniques lie the new fields of Computational Mathematical Modeling (CMM), including Computational Mechanics, Physics, Fluid Dynamics, Electromagnetics and Chemistry, all based on solving systems of differential equations using computers, combined with geometric modeling/Computer Aided Design (CAD). Computational modeling is also finding revolutionary new applications in biology, medicine, environmental sciences, economy and financial markets.

Education in mathematics forms the basis of science and engineering education from undergraduate to graduate level, because engineering and science are largely based on mathematical modeling. The level and the quality of mathematics education sets the level of the education as a whole.

The new technology of CMM/CAD crosses borders between traditional engineering disciplines and schools, and drives strong forces to modernize engineering education in both content and form from basic to graduate level.

Our Reform Program

Computational Mathematical → *Modeling* ←

Our own reform work started some 20 years ago in courses in CMM at advanced undergraduate level, and has through the years successively penetrated through the system to the basic education in calculus and linear algebra. Our aim has become to develop a complete program for mathematics education in science and engineering from basic undergraduate to graduate education. As of now our program contains the series of books:

1. *Computational Differential Equations*, (*CDE*)

2. *Applied Mathematics: Body & Soul I-III*, (*AM I-III*)

3. *Applied Mathematics: Body & Soul VI-*, (*AM IV-*).

AM I-III is the present book in three volumes I-III covering the basics of calculus and linear algebra. *AM IV-* offers a continuation with a series of volumes dedicated to specific areas of applications such as *Dynamical Systems (IV)*, *Fluid Mechanics (V)*, *Solid Mechanics (VI)* and *Electromagnetics (VII)*, which will start appearing in 2003. *CDE* published in 1996 may be be viewed as a first version of the whole *Applied Mathematics: Body & Soul* project.

Our program also contains a variety of software (collected in the *Mathematics Laboratory*), and complementary material with step-by step instructions for self-study, problems with solutions, and projects, all freely available on-line from the web site of the book. Our ambition is to offer a "box" containing a set of books, software and additional instructional material, which can serve as a basis for a full applied mathematics program in science and engineering from basic to graduate level. Of course, we hope this to be an on-going project with new material being added gradually.

We have been running an applied mathematics program based on *AM I-III* from first year for the students of chemical engineering at Chalmers since the Fall 99, and we have used parts of the material from *AM IV-* in advanced undergraduate/beginning graduate courses.

Main Features of the Program:

- The program is based on a synthesis of mathematics, computation and application.

- The program is based on new literature, giving a new unified presentation from the start based on constructive mathematical methods including a computational methodology for differential equations. *Whoopee!*

- The program contains, as an integrated part, software at different levels of complexity.

- The student acquires solid skills of implementing computational methods and developing applications and software using MATLAB. *OK, MATLAB.*

- The synthesis of mathematics and computation opens mathematics education to applications, and gives a basis for the effective use of modern mathematical methods in mechanics, physics, chemistry and applied subjects.

- The synthesis building on constructive mathematics gives a synergetic effect allowing the study of complex systems already in the basic education, including the basic models of mechanical systems, heat conduction, wave propagation, elasticity, fluid flow, electro-magnetism, reaction-diffusion, molecular dynamics, as well as corresponding multi-physics problems. *Ooops! Synergetic!*

- The program increases the motivation of the student by applying mathematical methods to interesting and important concrete problems already from the start.

- Emphasis may be put on problem solving, project work and presentation.

- The program gives theoretical and computational tools and builds confidence.

- The program contains most of the traditional material from basic courses in analysis and linear algebra

- The program includes much material often left out in traditional programs such as constructive proofs of all the basic theorems in analysis and linear algebra and advanced topics such as nonlinear systems of algebraic/differential equations. *Really?!! Right on!*

- Emphasis is put on giving the student a solid understanding of basic mathematical concepts such as real numbers, Cauchy sequences, Lipschitz continuity, and constructive tools for solving algebraic/differential equations, together with an ability to utilize these tools in advanced applications such as molecular dynamics.

- The program may be run at different levels of ambition concerning both mathematical analysis and computation, while keeping a common basic core.

Can probably use Mathematica too.

This is the first time this word has appeared.

AM I-III in Brief

Roughly speaking, *AM I-III* contains a synthesis of calculus and linear algebra including computational methods and a variety of applications. Emphasis is put on constructive/computational methods with the double aim of making the mathematics both understandable and useful. Our ambition is to introduce the student early (from the perspective of traditional education) to both advanced mathematical concepts (such as Lipschitz continuity, Cauchy sequence, contraction mapping, initial-value problem for systems of differential equations) and advanced applications such as Lagrangian mechanics, n-body systems, population models, elasticity and electrical circuits, with an approach based on constructive/computational methods.

Thus the idea is that making the student comfortable with both advanced mathematical concepts and modern computational techniques, will open a wealth of possibilities of applying mathematics to problems of real interest. This is in contrast to traditional education where the emphasis is usually put on a set of analytical techniques within a conceptual framework of more limited scope. For example: we already lead the student in the second quarter to write (in MATLAB) his/her own solver for general systems of ordinary differential equations based on mathematically sound principles (high conceptual and computational level), while traditional education at the same time often focuses on training the student to master a bag of tricks for symbolic integration. We also teach the student some tricks to that purpose, but our overall goal is different.

[handwritten margin note, left: A bag of symbolic tricks, indeed.]

Constructive Mathematics: Body & Soul

In our work we have been led to the conviction that the constructive aspects of calculus and linear algebra need to be strengthened. Of course, constructive and computational mathematics are closely related and the development of the computer has boosted computational mathematics in recent years. Mathematical modeling has two basic dual aspects: one symbolic and the other constructive-numerical, which reflect the duality between the infinite and the finite, or the continuous and the discrete. The two aspects have been closely intertwined throughout the development of modern science from the development of calculus in the work of Euler, Lagrange, Laplace and Gauss into the work of von Neumann in our time. For example, Laplace's monumental *Mécanique Céleste* in five volumes presents a symbolic calculus for a mathematical model of gravitation taking the form of Laplace's equation, together with massive numerical computations giving concrete information concerning the motion of the planets in our solar system.

[handwritten margin note, right: I like this duality.]

However, beginning with the search for rigor in the foundations of calculus in the 19th century, a split between the symbolic and constructive aspects gradually developed. The split accelerated with the invention of the electronic computer in the 1940s, after which the constructive aspects were pursued in the new fields of numerical analysis and computing sciences, primarily developed outside departments of mathematics. The unfortunate result today is that symbolic mathematics and constructive-numerical mathematics by and large are separate disciplines and are rarely taught together. Typically, a student first meets calculus restricted to its symbolic form and then much later, in a different context, is confronted with the computational side. This state of affairs lacks a sound scientific motivation and causes severe difficulties in courses in physics, mechanics and applied sciences which build on mathematical modeling.

Send quote to Eva?

New possibilities are opened by creating from the start a synthesis of constructive and symbolic mathematics representing a synthesis of Body & Soul: with computational techniques available the students may become familiar with nonlinear systems of differential equations already in early calculus, with a wealth of applications. Another consequence is that the basics of calculus, including concepts like real number, Cauchy sequence, convergence, fixed point iteration, contraction mapping, is lifted out of the wardrobe of mathematical obscurities into the real world with direct practical importance. In one shot one can make mathematics education both deeper and broader and lift it to a higher level. This idea underlies the present book, which thus in the setting of a standard engineering program, contains all the basic theorems of calculus including the proofs normally taught only in special honors courses, together with advanced applications such as systems of nonlinear differential equations. We have found that this seemingly impossible program indeed works surprisingly well. Admittedly, this is hard to believe without making real life experiments. We hope the reader will feel encouraged to do so.

Proofs and Theorems

Most mathematics books including Calculus texts follow a theorem-proof style, where first a theorem is presented and then a corresponding proof is given. This is seldom appreciated very much by the students, who often have difficulties with the role and nature of the proof concept.

We usually turn this around and first present a line of thought leading to some result, and then we state a corresponding theorem as a summary of the hypothesis and the main result obtained. We thus rather use a proof-theorem format. We believe this is in fact often more natural than the theorem-proof style, since by first presenting the line of thought the different ingredients, like hypotheses, may be introduced in a logical order. The

This is much better.

proof will then be just like any other line of thought, where one successively derives consequences from some starting point using different hypothesis as one goes along. We hope this will help to eliminate the often perceived mystery of proofs, simply because the student will not be aware of the fact that a proof is being presented; it will just be a logical line of thought, like any logical line of thought in everyday life. Only when the line of thought is finished, one may go back and call it a proof, and in a theorem collect the main result arrived at, including the required hypotheses. As a consequence, in the Latex version of the book we do use a theorem-environment, but not any proof-environment; the proof is just a logical line of thought preceding a theorem collecting the hypothesis and the main result.

They use Latex? Evidently.

The Mathematics Laboratory

We have developed various pieces of software to support our program into what we refer to as the *Mathematics Laboratory*. Some of the software serves the purpose of illustrating mathematical concepts such as roots of equations, Lipschitz continuity, fixed point iteration, differentiability, the definition of the integral and basic calculus for functions of several variables; other pieces are supposed to be used as models for the students own computer realizations; finally some pieces are aimed at applications such as solvers for differential equations. New pieces are being added continuously. Our ambition is to also add different multi-media realizations of various parts of the material.

In our program the students get a training from start in using MATLAB as a tool for computation. The development of the constructive mathematical aspects of the basic topics of real numbers, functions, equations, derivatives and integrals, goes hand in hand with experience of solving equations with fixed point iteration or Newton's method, quadrature, and numerical methods or differential equations. The students see from their own experience that abstract symbolic concepts have roots deep down into constructive computation, which also gives a direct coupling to applications and physical reality.

Go to http://www.phi.chalmers.se/bodysoul/

The *Applied Mathematics: Body & Soul* project has a web site containing additional instructional material and the *Mathematics Laboratory*. We hope that the web site for the student will be a good friend helping to (independently) digest and progress through the material, and that for the teacher it may offer inspiration. We also hope the web site may serve as a forum for exchange of ideas and experience related the project, and we therefore invite both students and teachers to submit material.

> My heart is sad and lonely
> for you I sigh, dear, only
> Why haven't you seen it
> I'm all for you body and soul
> (Green, Body and Soul)

Appendix C
Public Debate

Get this (economic plan) passed. Later on, we can all debate it.
(President George Bush, to New Hampshire legislators)

There is some public debate on mathematics education in Sweden, and probably also in other countries. There seems to be a common agreement that there is a crisis in mathematics education today (although the chairman of the Mathematics Delegation does not believe there is). A common view among professional mathematicians teaching mathematics at a university/college, is that the main manifestation of the the crisis is that the students entering the university/college have an inadequate training in mathematics from high-school. This represents a *privilege of problem definition*. This term was coined by the famous Swedish author Lars Gustafsson, since many years active from a platform outside Sweden at the University of Austin, during the Marxist heydays of 1968. It is clear that to have (or take) this privilege gives an advantage in a debate. But of course, it is risky, since if the privilege is lost, the debate may be lost as well.

The problem of mathematics education at the university/college, and apparently there is a problem, is thus identified to be caused by a lack of training in high-school; the students know too little mathematics when they enter the university, which effectively prevents them from learning more. This message has been delivered in several debate articles in the Swedish press, usually signed by a group of university professors of mathematics. We refer to this as the *standard problem definition*.

In several debate articles in the Swedish press, we have questioned the standard problem definition, with arguments like those presented in this book. We have received quite a bit of positive response from many, but none from the (many) professors behind the standard problem definition. We present below our latest debate article (published on March 6 2004 in *Göteborgs-Posten GP*, the second paper in Sweden with 600.000 copies/day), where we put forward the idea that the lack of response from professional mathematicians may come from a lack of training in computational mathematics. Notably, there was no response to this article. Maybe Wittgenstein's famous "Whereof one cannot speak, thereof one must be silent." describes the situation.

Further, teachers at high-schools say nothing to defend themselves, because in the established hierarchy of science they cannot argue with university professors. The situation is similar for the experts of mathematics didactics in charge of the education of high-school teachers. The net result is that the debate, in the sense of exchange of ideas and viewpoints, is almost non-existent.

How true!

Our debate article reads as follows, with the head lines set by the newspaper.

Old Fashioned View of Mathematics Behind Crisis

A recent debate article in GP signed by a group of university professors sends yet another message that mathematics education is in a state of crisis. The next day the article is followed up by a report from the mathematics education at Chalmers where many students get bogged down. (What a waste of human resources!).

The idea that mathematics education on all levels is in a state of crisis, is widely spread and made Östros form the Mathematics Delegation. Many university professors claim that the root of the crisis at the university is the inadequate training the students get in high school mathematics, as in the debate article and the report.

A Deeper Reason?

But is there some deeper reason behind the crisis, so vividly witnessed by so many? Could it be that the main problem is not so much the student, but instead the professor? Could it be that the reason is that mathematics as a religion with its priests is in a state of crisis, and that the congregation therefore is loosing faith? What a heretical thought! It can't be true, or can it?

Yes, it can. Such things happen in science, and mathematics is a science. New views replace older ones in changes of paradigm, or scientific revolutions, usually during extended periods of agony, according to the famous Thomas Kuhn, who has carefully analyzed the phenomenon.

Such a scientific revolution is now going on in mathematics caused by the new possibilities of mathematical computation the computer offers. Our new Information Society with word, image and sound in digital form is based on computational mathematics, which is the mathematics used by the computer. The computer is changing mathematics as a science and as a tool in our society.

Today we may speak of mathematics without computer, as presented in the traditional mathematics education at Chalmers, with analytical formulas as the kernel (see the report), and mathematics with computer, which is the new computational mathematics with algorithms for computation as the kernel.

There is no sharp line between these areas: an algorithm is expressed in analytical mathematical formulas, which are translated into computer code, and good formulas may give understanding and insight. The mathematics professors who signed the debate article and express their views in the report all represent mathematics without computer, and seem to share the view that the root of the crisis at the university is inadequate training in high-school.

Modern Education

What would a modern mathematics education (mathematics with computer) look like? Is there anything like that? Yes! We have developed a reformed mathematics education, which is e.g. used for the students of chemical engineering at Chalmers. We refer to our reform program as Body&Soul, where Body represents computation and Soul mathematical analysis. And indeed, the students of chemical engineering manage their studies in mathematics quite well, despite the facts that our program has a high level and mathematics is difficult. Many students are surprisingly good, and very few fail completely.

We believe that we have shown a route to a both understandable and useful modern education. We continue our work towards a reformed high-school education, following the principle that the top brick in the building of knowledge has to be put according to the current standpoint of science, and the bricks below so as to give support.

Lack of Debate

One could now expect the debate about our program to be lively, if it could contribute to resolve the crisis? But there is no debate. We know this from making several attempts in GP and Dagens Nyheter, which have all been met with silence. We believe the reason may be that expert knowledge in computational mathematics is required to constructively debate (the top brick of a) modern mathematics education, and we don't see this knowledge among the professors behind the debate article and the report, neither among all the members of the Mathematics Delegation. Maybe in this context one could speak of inadequate training?

Who would be able to say that what we are saying is not correct? Who would dare to say that what we are saying is maybe quite true?

Seek Answers

But the problem remains: How to come out of the crisis if you don't want to debate? If you don't want to reform? If professional mathematicians can't agree on where mathematics as a science stands today. If you want to focus on inadequate training in high-school only, and not scrutinize your own role and the content and form of the education at the university? If the Mathematics Delegation refuses to engage expertise of computational mathematics and does not want to seek answers to questions of what and why?

Kenneth Eriksson, Prof of Applied Mathematics, College of Trollhättan,
Johan Hoffman, Post Doc, Courant Inst of Mathematical Sciences, NY,
Claes Johnson, Prof of Applied Mathematics, Chalmers,
Anders Logg, Ph D student, Computational Mathematics, Chalmers,
Nils Svanstedt, Prof of Mathematics, University of Göteborg.

24 Oct 2005